The Human Brain during the First Trimester 57- to 60-mm Crown-Rump Lengths

This seventh of 15 short atlases reimagines the classic 5-volume *Atlas of Human Central Nervous System Development*. This volume presents serial sections from specimens between 57 mm and 60 mm with detailed annotations, together with 3D reconstructions. An introduction summarizes human CNS development by using high-resolution photos of methacrylate-embedded rat embryos at a similar stage of development as the human specimens in this volume. The accompanying Glossary gives definitions for all the terms used in this volume and all the others in the *Atlas*.

Features

- Classic anatomical atlas
- Detailed labeling of structures in the developing brain offers updated terminology and the identification of unique developmental features, such as germinal matrices of specific neuronal populations and migratory streams of young neurons
- Appeals to neuroanatomists, developmental biologists, and clinical practitioners
- A valuable reference work on brain development that will be relevant for decades

ATLAS OF
HUMAN CENTRAL NERVOUS SYSTEM DEVELOPMENT
Series

Volume 1: The Human Brain during the First Trimester 3.5- to 4.5-mm Crown-Rump Lengths

Volume 2: The Human Brain during the First Trimester 6.3- to 10.5-mm Crown-Rump Lengths

Volume 3: The Human Brain during the First Trimester 15- to 18-mm Crown-Rump Lengths

Volume 4: The Human Brain during the First Trimester 21- to 23-mm Crown-Rump Lengths

Volume 5: The Human Brain during the First Trimester 31- to 33-mm Crown-Rump Lengths

Volume 6: The Human Brain during the First Trimester 40- to 42-mm Crown-Rump Lengths

Volume 7: The Human Brain during the First Trimester 57- to 60-mm Crown-Rump Lengths

Volume 8: The Human Brain during the Second Trimester 96- to 150-mm Crown-Rump Lengths

Volume 9: The Human Brain during the Second Trimester 160- to 170-mm Crown-Rump Lengths

Volume 10: The Human Brain during the Second Trimester 190- to 210-mm Crown-Rump Lengths

Volume 11: The Human Brain during the Third Trimester 225- to 235-mm Crown-Rump Lengths

Volume 12: The Human Brain during the Third Trimester 260- to 270-mm Crown-Rump Lengths

Volume 13: The Human Brain during the Third Trimester 310- to 350-mm Crown-Rump Lengths

Volume 14: The Spinal Cord during the First Trimester

Volume 15: The Spinal Cord during the Second and Third Trimesters and the Early Postnatal Period

The Human Brain during the First Trimester 57- to 60-mm Crown-Rump Lengths

Atlas of Human Central Nervous System Development, Volume 7

Shirley A. Bayer
Joseph Altman

CRC Press
Taylor & Francis Group
Boca Raton London New York

CRC Press is an imprint of the
Taylor & Francis Group, an **informa** business

First edition published 2023
by CRC Press
6000 Broken Sound Parkway NW, Suite 300, Boca Raton, FL 33487-2742

and by CRC Press
4 Park Square, Milton Park, Abingdon, Oxon, OX14 4RN

CRC Press is an imprint of Taylor & Francis Group, LLC

LCCN 2022008216

ISBN: 978-1-032-18335-0 (hbk)
ISBN: 978-1-032-18566-8 (pbk)
ISBN: 978-1-003-27066-9 (ebk)

DOI: 10.1201/9781003270669

Publisher's note: This book has been prepared from camera-ready copy provided by the authors.

Access the Support Material: www.routledge.com/9781032183350

Typeset in Times Roman by KnowledgeWorks Global Ltd.

CONTENTS

ACKNOWLEDGMENTS --vii

AUTHORS-- ix

PART I.   **INTRODUCTION** ---1

Organization of the Atlas --1

Plate Preparation ---1

Development in Specimens (CR 57 to 60 mm) ---------------------------------2

References --- 11

PART II.   **57-mm Crown-Rump Length, C1500** --------------------------------------- **12**

Low-Magnification Plates 1–21 A–B ---14–**55**

PART III.   **60-mm Crown-Rump Length, Y1-59** -- **56**

Low-Magnification Plates 22–42 A–B ---------------------------------------58–99

High-Magnification Plates 43–56 A–B---100–127

ACKNOWLEDGMENTS

We thank the late Dr. William DeMyer, pediatric neurologist at Indiana University Medical Center, for access to his personal library on human CNS development. We also thank the staff of the National Museum of Health and Medicine that were at the Armed Forces Institute of Pathology, Walter Reed Hospital, Washington, D.C. when we collected data in 1995 and 1996: Dr. Adrianne Noe, Director; Archibald J. Fobbs, Curator of the Yakovlev Collection; Elizabeth C. Lockett; and William Discher. We are most grateful to the late Dr. James M. Petras at the Walter Reed Institute of Research who made his darkroom facilities available so that we could develop all the photomicrographs on location rather than in our laboratory in Indiana. Finally, we thank Chuck Crumly, Neha Bhatt, Kara Roberts, Michele Dimont, and Rebecca Condit for expert help during production of the manuscript.

AUTHORS

Shirley A. Bayer received her PhD from Purdue University in 1974 and spent most of her scientific career working with Joseph Altman. She was a professor of biology at Indiana-Purdue University in Indianapolis for several years, where she taught courses in human anatomy and developmental neurobiology while continuing to do research in brain development. Her lengthy publication record of dozens of peer-reviewed, scientific journal articles extends back to the mid 1970s. She has co-authored several books and many articles with her late spouse, Joseph Altman. It was her research (published in *Science* in 1982) that proved that new neurons are added to granule cells in the dentate gyrus during adult life, a unique neuronal population that grows. That paper stimulated interest in the dormant field of adult neurogenesis.

Joseph Altman, now deceased, was born in Hungary and migrated with his family via Germany and Australia to the US. In New York, he became a graduate student in psychology in the laboratory of Hans-Lukas Teuber, earning a PhD in 1959 from New York University. He was a postdoctoral fellow at Columbia University, and later joined the faculty at the Massachusetts Institute of Technology. In 1968, he accepted a position as a professor of biology at Purdue University. During his career, he collaborated closely with Shirley A. Bayer. From the early 1960s-2016, he published many articles in peer-reviewed journals, books, monographs, and free online books that emphasized developmental processes in brain anatomy and function. His most important discovery was adult neurogenesis, the creation of new neurons in the adult brain. This discovery was made in the early 1960s while he was based at MIT, but was largely ignored in favor of the prevailing dogma that neurogenesis is limited to prenatal development. After Dr. Bayer's paper proved new neurons are added to granule cells in the hippocampus, Dr. Altman's monumental discovery became more accepted. During the 1990s, new researchers "rediscovered" and confirmed his original finding. Adult neurogenesis has recently been proven to occur in the dentate gyrus, olfactory bulb, and striatum through the measurement of Carbon-14—the levels of which changed during nuclear bomb testing throughout the 20th century—in postmortem human brains. Today, many laboratories around the world are continuing to study the importance of adult neurogenesis in brain function. In 2011, Dr. Altman was awarded the Prince of Asturias Award, an annual prize given in Spain by the Prince of Asturias Foundation to individuals, entities, or organizations globally who make notable achievements in the sciences, humanities, and public affairs. In 2012, he received the International Prize for Biology - an annual award from the Japan Society for the Promotion of Science (JSPS) for "outstanding contribution to the advancement of research in fundamental biology." This Prize is one of the most prestigious honors a scientist can receive. When Dr. Altman died in 2016, Dr. Bayer continued the work they started over 50 years ago. In her late husband's honor, she created the Altman Prize, awarded each year by JSPS to an outstanding young researcher in developmental neuroscience.

INTRODUCTION

ORGANIZATION OF THE ATLAS

This is the seventh book in the *Atlas of Human Central Nervous System Development* series, 2nd Edition. It deals with human brain development in two normal specimens at the end of the first trimester with crown-rump (CR) lengths from 57 to 60 mm and estimated gestation weeks (GW) from 11.9 to 12.5 (Loughna et al., 2009). These specimens were analyzed in Volume 4 of the 1st Edition (Bayer and Altman, 2006). One specimen (Y-159) is from the *Yakovlev Collection.*[1] The other specimen (C1500) is from the *Carnegie Collection.*[2] Both collections are in the National Museum of Health and Medicine, which used to be located at the Armed Forces Institute of Pathology (AFIP) in Walter Reed Hospital in Washington, D.C. When the AFIP closed, the National Museum moved to Silver Springs, MD; this Collection is still available for research. Y1-59 is cut in the frontal plane; C1500 in the horizontal plane. Unfortunately, there is no sagitally-sectioned specimen available at the same stage of development. As in the previous volumes of the *Atlas*, each specimen is presented in serial gray-scale photographs of its sections showing the brain and surrounding tissues (**Parts II–III**). The photographs are shown from anterior to posterior (frontal specimen) and dorsal to ventral (horizontal specimen). The dorsal part of each frontal section is toward the top of the page, the ventral part at the bottom, and the midline is in the vertical center. In the horizontal sections, the left side is anterior, right side, posterior, and the midline is in the horizontal center.

PLATE PREPARATION

All sections of a given specimen were photographed at the same magnification. Sections throughout the entire specimen were photographed in serial order with Kodak technical pan black-and-white negative film (#TP442). The film was developed for 6 to 7 minutes in dilution F of Kodak HC-110 developer, stop bath for 30 seconds, Kodak fixer for 5 minutes, Kodak hypo-clearing agent for 1 minute, running water rinse for 10 minutes, and a brief rinse in Kodak photo-flo before drying. The negatives were scanned at 2700 dots per inch (dpi) as color positives with a Nikon Coolscan-1000 35-mm negative film scanner attached to a Macintosh PowerMac G3 computer which had a plug-in driver built into Adobe Photoshop. The original scans were converted to 300 dpi using the non-resampling method for image size. The powerful features of Adobe photoshop were used to enhance contrast, correct uneven staining, and slightly darken or lighten areas of uneven exposure.

1. The *Yakovlev Collection* (designated by a **Y** prefix in the specimen number) is the work of Dr. Paul Ivan Yakovlev (1894–1983), a neurologist affiliated with Harvard University and the AFIP. Throughout his career, Yakovlev collected many diseased and normal human brains. He invented a giant microtome that was capable of sectioning entire human brains. Later, he became interested in the developing brain and collected many human brains during the second and third trimesters. The normal brains in the developmental group were cataloged by Haleem (1990) and were examined by us during 1996 and 1997 when we spent time at the AFIP.

2. The *Carnegie Collection* (designated by a **C** prefix in the specimen number) started in the Department of Embryology of the Carnegie Institution of Washington. It was led by Franklin P. Mall (1862–1917), George L. Streeter (1873–1948), and George W. Corner (1889–1981). These specimens were collected during a span of 40 to 50 years and were histologically prepared with a variety of fixatives, embedding media, cutting planes, and histological stains. Early analyses of specimens were published in the early 1900s in *Contributions to Embryology, The Carnegie Institute of Washington* (now archived in the Smithsonian Libraries). O'Rahilly and Müller (1987, 1994) have given overviews of some first trimester specimens in this collection.

The photos chosen for annotation in **Parts II–III** are presented as companion plates. The *low-magnification plates* of both specimens are designated as **A** and **B** on facing pages. **Part A** on the left shows the full-contrast photo, while **Part B** on the right shows a low-contrast copy with annotations. A few *high-magnification plates* of the frontal specimen feature enlarged views of the brain to show tissue organization. This allows users to see the entire section and then consult the detailed markup in the low-contrast copy on the facing page, leaving little doubt about what is being identified. The labels themselves are not abbreviated, so there is no lookup on a list. Different fonts are used to label different classes of structures: the ventricular system is labeled in **CAPITALS**, the neuroepithelium and other germinal zones in **Helvetica bold**, transient structures in ***Times bold italic***, and permanent structures in Times Roman or **Times bold**. Adobe Illustrator was used to superimpose labels and to outline structural details on the low-contrast images. The plates were placed into a book layout using Adobe InDesign. Finally, high-resolution portable document files (PDFs) were uploaded to CRC Press/Taylor & Francis websites.

DEVELOPMENT IN SPECIMENS
(CR 57–60 mm)

The specimens in this volume are equivalent to rat embryos on embryonic day (E) 19 based on our developmental-stage matching. Our timetables of neurogenesis use ^{3}H-thymidine dating methods (Bayer and Altman, 1991, 1995, 2012-present; Bayer et al., 1993, 1995) to determine neuronal populations that are being generated in E19 rats; we assume that is comparable to neurogenesis in 57 to 60-mm human specimens (Bayer et al., 1993, 1995; Bayer and Altman, 1995). **Table 1A** lists populations being generated from the medulla to the diencephalon. **Table 1B** lists populations being generated in the telencephalon. We use photos of a rat embryo on E19 (Bayer, 2013-present) that was exposed to ^{3}H-thymidine 2 hours before death to show details of the few active germinal zones throughout the brain (**Figs. 1–7**).

In the medulla, pontine nuclear neurons are still being generated in the precerebellar neuroepithelium (NEP, **Fig. 2**). In the cerebellum, the external germinal layer (egl) in lateral (**Fig. 2**) and medial (**Fig. 3**) parts continues to grow over the cerebellar cortex while Purkinje neurons migrate upward to settle beneath it. Many deep nuclear neurons are leaving superficial parts to settle beneath the growing cortex. The egl is one prong of a germinal trigone. The only active germinal zone in the midbrain is the medial part of the inferior colliculus (**Fig. 3**) where neurons are being generated for anterior and posterior medial parts and anterior intermediate parts. In the hypothalamus, we illustrate the active germinal zone of the arcuate nucleus (**Fig. 4**). In the telencephalon, we show details of the germinal zones in the neocortex (**Fig. 5**), the hippocampus (**Fig. 6**), and the complex germinal zones of the anterobasal forebrain (**Fig. 7**).

Table 1A: Neurogenesis by Region	
REGION and NEURAL POPULATION	CROWN-RUMP LENGTH 57–60 mm
PRECEREBELLAR NUCLEI	
Pontine nuclei	●
CEREBELLAR CORTEX	
Golgi cells	●
INFERIOR COLLICULUS	
Posterior intermediate	● ●
Anteromedial	● ●
Posteromedial	● ●
EPITHALAMUS	
Medial habenula	●
PREOPTIC AREA/HYPOTHALAMUS	
Medial preoptic nucleus	●
Sexually dimorphic nucleus	● ●
Arcuate nucleus	●

Table 1A. Neural populations being generated from the medulla through the diencephalon in rats (comparable to humans at CR 57–60 mm). *Green dots* indicate the amount of neurogenesis occurring: one dot=<15%; two dots=15-90%. This same dot notation is used for **Table 1B**.

Table 1B: Neurogenesis by Region	
REGION and NEURAL POPULATION	CROWN-RUMP LENGTH 57–60 mm
PALLIDUM AND STRIATUM	
Olfactory tubercle (small neurons)	● ●
Caudate and putamen	● ●
Nucleus accumbens	● ●
Islands of Calleja	● ●
AMYGDALA	
Intercalated masses	●
Amygdalo-hippocampal area	● ●
NEOCORTEX and LIMBIC CORTEX	
Layera IV-II	● ●
OLFACTORY CORTEX	
Layer II (anterior)	●
HIPPOCAMPAL REGION	
Subiculum (superficial)	● ●
Ammon's Horn CA1	● ●
Ammon's Horn CA3	● ●
Ammon's Horn CA4	● ●
OLFACTORY BULB	
External tufted cells (main bulb)	● ●
ANTERIOR OLFACTORY NUCLEUS	
Pars externa	●
AON proper	● ●

GERMINAL ZONES (covered with black dots) IN AN E19 RAT EMBRYO EXPOSED TO ³H-THYMIDINE 2 HOURS BEFORE DEATH

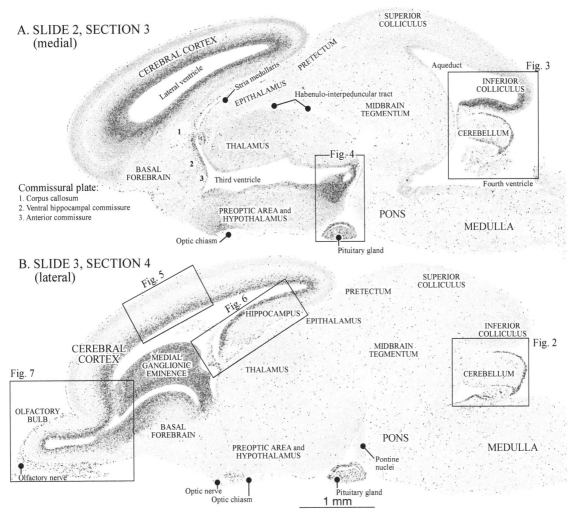

Figure 1. Medial (**A**) and slightly more lateral (**B**) sagittal sections of the brain in an E19 rat embryo at a similar stage of development as the human specimens in this volume. ³H-thymidine was injected into the mother 2 hrs before she was killed and this embryo was removed. The head was immersed in Bouin's fixative and embedded in paraffin. Serial 6-μm sections were placed on microscopic slides, covered with photographic emulsion, and were developed after a 6-week exposure period. The sections were post-stained with hematoxylin. ³H-thymidine will only be absorbed by cells in the S-phase of the mitotic cycle, so the labeled areas (covered with black dots) only mark the germinal zones. The *boxes* indicate the areas shown in greater detail in **Figures 2–7**. Source: braindevelopmentmaps.org (E19 2hr survival archive)

Since these specimens are at the end of the first trimester, it is appropriate to review some developmental highlights. The first trimester is the time during which the most profound and dramatic changes take place. In terms of the central nervous system, it is the time when the "floorplan" originates and comes to fruition. We started with a sheet of pseudostratified columnar epithelium, the neuroepithelium, that is draped over the developing body. This tissue is responsive to the cues coming from that body and is mosaicized to generate unique neuronal populations from the most anterior forebrain to the posterior tip of the spinal cord. At specific times, the neuroepithelium produces neurons in a highly choreographed manner so that neurons leave their germinal zones, migrate (some leaving behind axons), and settle in the parenchyma *at the right time and in the right place to form functional circuits in the maturing and adult brain.*

The parenchyma continually expands during the first trimester while most of the neuroepithelium eventually disappears by the end of the first trimester. The only germinal zones surviving into the second trimester (and some into the third trimester and even throughout life) are the cerebellar external germinal layer, the cerebral cortical neuroepithelium and subventricular zone, the striatal subventricular zone, the hippocampal neuroepithelium, the dentate subgranular zone, and the rostral migratory stream in olfactory areas.

CEREBELLUM and PRECEREBELLAR NEUROEPITHELIUM (NEP)

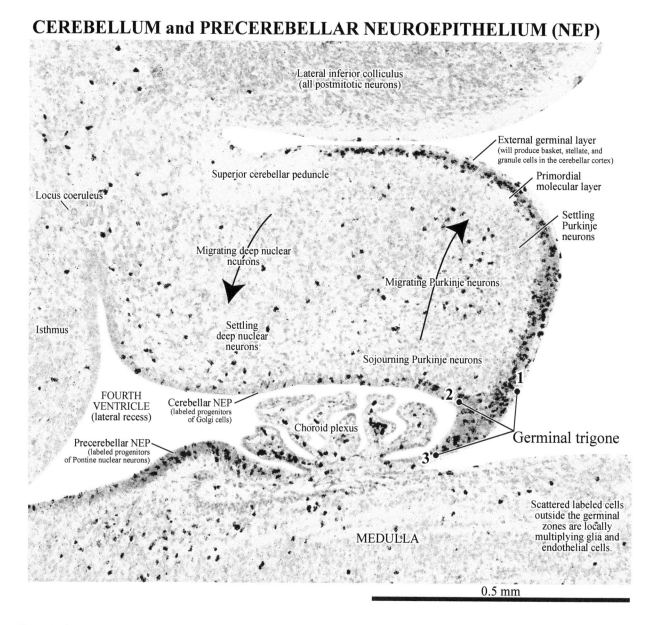

Figure 2. The cerebellum and precerebellar neuroepithelium from the boxed area in **Figure 1B**. Source: braindevelopmentmaps.org (E19 2hr survival archive)

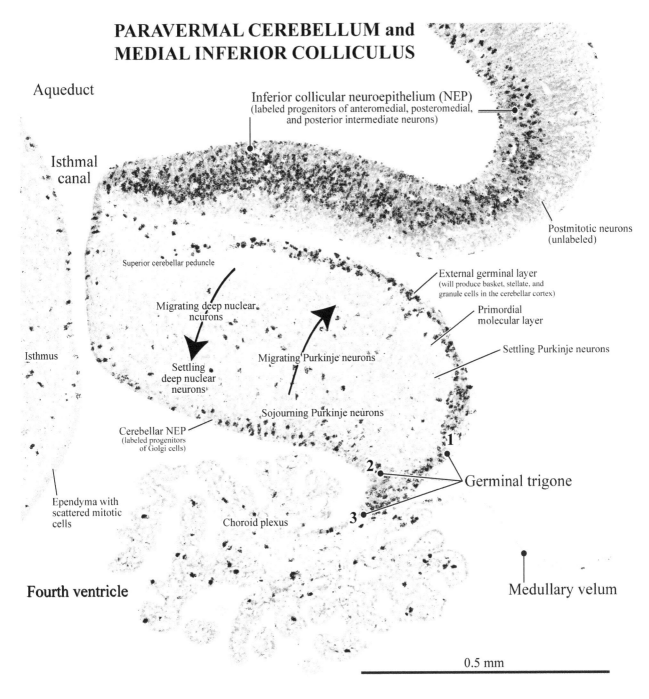

PARAVERMAL CEREBELLUM and MEDIAL INFERIOR COLLICULUS

Aqueduct

Inferior collicular neuroepithelium (NEP)
(labeled progenitors of anteromedial, posteromedial,
and posterior intermediate neurons)

Isthmal canal

Postmitotic neurons
(unlabeled)

Superior cerebellar peduncle

External germinal layer
(will produce basket, stellate, and
granule cells in the cerebellar cortex)

Migrating deep nuclear neurons

Primordial molecular layer

Isthmus

Settling deep nuclear neurons

Migrating Purkinje neurons

Settling Purkinje neurons

Sojourning Purkinje neurons

Cerebellar NEP
(labeled progenitors
of Golgi cells)

1

2

Germinal trigone

Ependyma with scattered mitotic cells

Choroid plexus

3

Fourth ventricle

Medullary velum

0.5 mm

Figure 3. The paravermal cerebellum and medial inferior colliculus from the boxed area in **Figure 1A**. Source: braindevelopmentmaps.org (E19 2hr survival archive)

6

HYPOTHALAMUS and PITUITARY GLAND

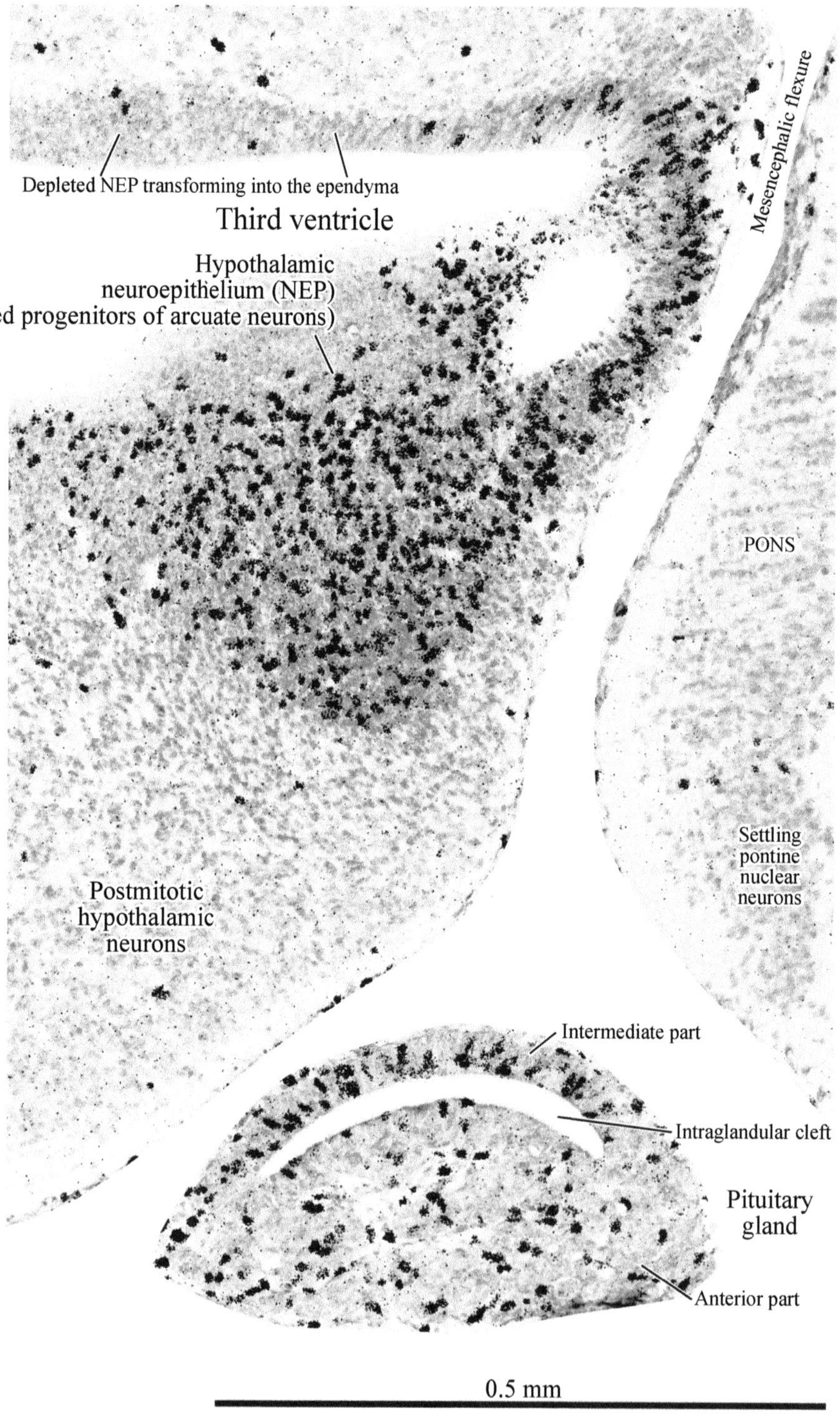

Depleted NEP transforming into the ependyma

Third ventricle

Hypothalamic
neuroepithelium (NEP)
(labeled progenitors of arcuate neurons)

Mesencephalic flexure

PONS

Settling
pontine
nuclear
neurons

Postmitotic
hypothalamic
neurons

Intermediate part

Intraglandular cleft

Pituitary
gland

Anterior part

0.5 mm

Figure 4. The germinal zone of the arcuate nucleus and the pituitary gland from the boxed area in **Figure 1A**. Source: braindevelopmentmaps.org (E19 2hr survival archive)

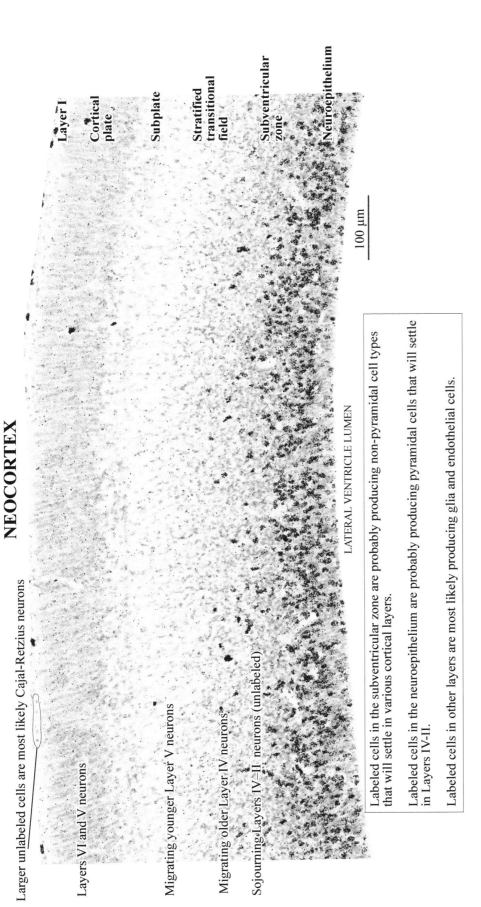

Figure 5. The neocortex from the boxed area in **Figure 1B**. Source: braindevelopmentmaps.org (E19 2hr survival archive)

8

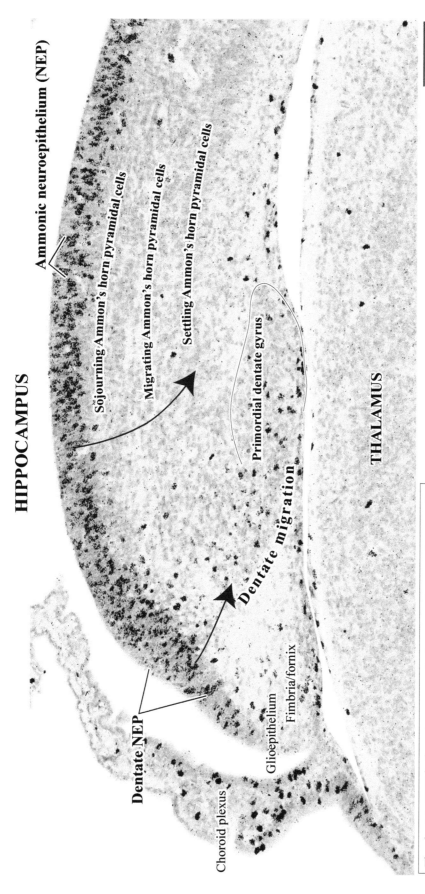

HIPPOCAMPUS

Ammonic neuroepithelium (NEP)

Sojourning Ammon's horn pyramidal cells

Migrating Ammon's horn pyramidal cells

Settling Ammon's horn pyramidal cells

Primordial dentate gyrus

Dentate migration

Dentate NEP

Choroid plexus

Glioepithelium

Fimbria/fornix

THALAMUS

The dentate NEP is the primary germinal source of a secondary matrix that migrates (and proliferates as it is migrating) into the primordial dentate gyrus. This will become the **subgranular zone** that generates granule cells throughout fetal, infant, juvenile, and adult life.

Figure 6. The hippocampus from the boxed area in **Figure 1B**. Source: braindevelopmentmaps.org (E19 2hr survival archive)

ANTEROBASAL TELENCEPHALON

The **rostral migratory stream** is a conduit of proliferating cells that is a source of granule cells in the olfactory bulb. It is continuous with the NEPs and SVZs of the cerebral cortex and the basal forebrain. In rats, the stream persists in juveniles and a remnant is retained in adults (Altman, 1969).

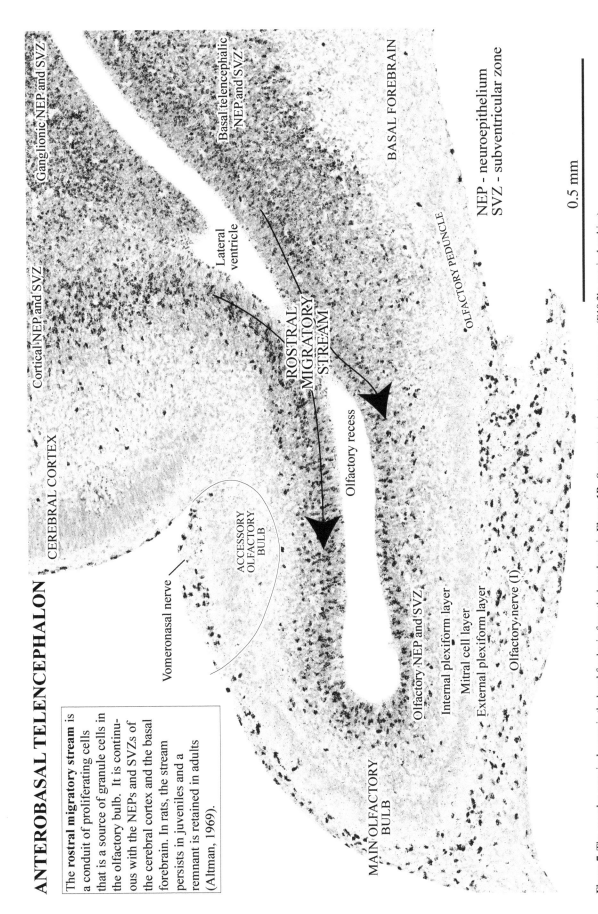

Ganglionic NEP and SVZ

Basal telencephalic NEP and SVZ

Cortical NEP and SVZ

BASAL FOREBRAIN

CEREBRAL CORTEX

Lateral ventricle

ROSTRAL MIGRATORY STREAM

Olfactory recess

OLFACTORY PEDUNCLE

NEP - neuroepithelium
SVZ - subventricular zone

0.5 mm

Vomeronasal nerve

ACCESSORY OLFACTORY BULB

MAIN OLFACTORY BULB

Olfactory NEP and SVZ

Internal plexiform layer

Mitral cell layer

External plexiform layer

Olfactory nerve (I)

Figure 7. The complex germinal zones in the basal forebrain from the boxed area in **Figure 1B**.   Source: braindevelopmentmaps.org (E19 2hr survival archive)

REFERENCES

Altman J (1969) Autoradiographic and histological studies of postnatal neurogenesis. IV. Cell proliferation and migration in the anterior forebrain, with special reference to persisting neurogenesis in the olfactory bulb. *Journal of Comparative Neurology*, 137:433-458.

Bayer, SA (in press) *Glossary to Accompany Atlas of Human Central Nervous System Development.* Taylor & Francis/CRC Press.

Bayer, SA (2013-present) www.braindevelopmentmaps.org Laboratory of Developmental Neurobiology, Ocala FL. (This website is an image database of methacrylate-embedded normal rat embryos and paraffin-embedded rat embryos exposed to ^{3}H-Thymidine.)

Bayer SA, Altman J (1991) *Neocortical Development*, Raven Press, New York.

Bayer SA, Altman J (1995) Development: Some principles of neurogenesis, neuronal migration and neural circuit formation. In: *The Rat Nervous System*, 2nd Edition, George Paxinos, Ed. Academic Press, Orlando, Florida, pp. 1079-1098.

Bayer SA, Altman J (1997) *Development of the Cerebellar System in Relation to Its Evolution, Structure, and Function*, CRC Press, Boca Raton, FL.

Bayer SA, Altman J (2006) *Atlas of Human Central Nervous System Development*, Volume 4: *The Human Brain during the Late First Trimester.* CRC Press.

Bayer SA, Altman J (2012-present) www.neurondevelopment.org (This website has downloadable pdf files of our scientific papers on rat brain development grouped by subject.)

Bayer SA, Altman J, Russo RJ, Zhang X (1993) Timetables of neurogenesis in the human brain based on experimentally determined patterns in the rat. *Neurotoxicology* **14**: 83-144.

Bayer SA, Altman J, Russo RJ, Zhang X (1995) Embryology. In: *Pediatric Neuropathology*, Serge Duckett, Ed. Williams and Wilkins, pp. 54-107.

Haleem M (1990) *Diagnostic Categories of the Yakovlev Collection of Normal and Pathological Anatomy and Development of the Brain*. Washington, D.C. Armed Forces Institute of Pathology.

Hochstetter F (1919) *Beiträge zur Entwicklungsgeschichte des menschlichen Gehirns*. Vol. 1. Leipzig und Wien: Deuticke.

Loughna P, Citty L, Evans T, Chudleigh T (2009) Fetal size and dating: Charts recommended for clinical obstetric practice, *Ultrasound*, 17:161-167.

O'Rahilly R, Müller F. (1987) *Developmental Stages in Human Embryos, Carnegie Institution of Washington*, Publication 637.

O'Rahilly R, Müller F. (1994) *The Embryonic Human Brain*, Wiley-Liss, New York.

PART II: C1500
CR 57 mm (GW 11.9)
Horizontal

This is specimen number 1500 in the Carnegie collection, designated here as C1500. A normal fetus with a crown-rump length (CR) of 57-mm was collected in 1916. The fetus is estimated to be in gestational week (GW) 11.9. The entire fetus was fixed in formalin and was cut in the horizontal plane. The records are not clear regarding section thickness. Since the brain is present in 2,400 sections,we estimate that the thickness is between 8- and 10-μm. All sections were stained with Bodian's method to show developing fiber tracts. There is no photograph of C1500's brain before it was embedded and cut, and a specimen from Hochstetter (1919) that is comparable in age is used to show the approximate section plane and external features at GW11.9 (**Figure 8**). Large sections containing the cerebral hemispheres, are shown at lower magnification in **Plates 1-13**. Small sections containing only the brainstem, are shown at a higher magnification in **Plates 14-21**. To maximize image size within page space, C1500's sections are rotated 90° (landscape orientation). The anterior part of each section is on the left (page bottom), and the posterior part of each section is on the right (page top).

Throughout the telencephalon, the neocortical neuroepithelium and subventricular zone are prominent. Indeed, the cortical primordium is now the largest structure in the brain. The stratified transitional field (STF) contains STF1 and STF5 throughout; with STF4 only in lateral areas. Both the STF layers and the cortical plate have a pronounced lateral (thicker) to medial (thinner) maturation gradient. The olfactory bulb beneath the anterior septum contains a small rostral migratory stream in its core. In anterolateral parts of the cerebral cortex, streams of neurons and glia appear to leave STF4 and enter the lateral migratory stream. The hippocampus is in an immature position dorsal to the thalamus and medial to the temporal lobe. Cells are entering Ammon's horn pyramidal layer in the ammonic migration, and granule cells and their precursors are migrating to the hilus of the presumptive dentate gyrus in the dentate migration; there is no granular layer. A massive neuroepithelium/subventricular zone overlies the nucleus accumbens and striatum (caudate and putamen) where neurons (and glia) are being generated.

The cerebellum is a thick, smooth plate overlying the posterior pons and medulla. However, there is still a thin neuroepithelium at the ventricular surface where Golgi cells are being generated; we postulate that all deep neurons and Purkinje cells have been generated earlier. The deep neurons are in place beneath the cortex, but have indefinite nuclear subdivisions. The cortical surface is covered by an external germinal layer (egl) that is actively producing neuronal stem cells, (for granule, stellate, and basket cells) that will later fill in the layers of the cerebellar cortex. Lamination in the cortex is nearly absent, except for a thin molecular layer beneath the egl. Nearly all Purkinje cells are migrating, and a few are settling.

Most of the third ventricle, aqueduct, and fourth ventricle are lined by a thin glioepithelium/ependyma indicating that neurogenesis in the primary neuroepithelium is complete. There are active neuroepithelial sites in the hypothalamus, inferior colliculus and in the precerebellar neuroepithelium.

Neurons throughout the diencephalon, midbrain tegmentum, pons, and medulla are settling. Because C1500 is not Nissl-stained, nuclear divisions are nearly invisible. The anterior extramural migratory stream is visible in the medulla and pons, and a fair number of neurons have already settled in the pontine nuclei.

The Bodian stain clearly shows several fiber tracts and nerves throughout the brainstem. The optic nerve and tract are well defined, along with the medial forebrain bundle. Pontine gray fibers cross the midline and a distinct middle cerebellar peduncle is present. There is also a distinct superior and inferior cerebellar peduncle. There is definite staining in the trigeminal nerve and tract, the facial nerve, the abducens nerve, and the glossopharyngeal nerve. In the telencephalon, many fibers are stained black in the internal capsule; some of these fibers are likely from the thalamus and head into the stratified transitional field to intermingle with migrating pyramidal cells in the cerebral cortex.

GW11.9 HORIZONTAL SECTION PLANES

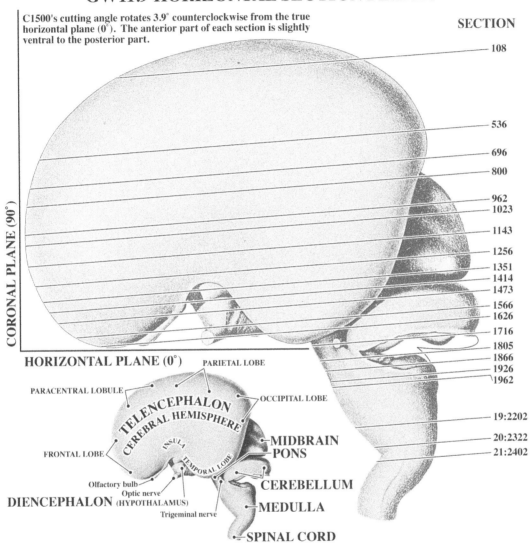

C1500's cutting angle rotates 3.9° counterclockwise from the true horizontal plane (0°). The anterior part of each section is slightly ventral to the posterior part.

SECTION

— 108

— 536

— 696
— 800

— 962
— 1023

— 1143

— 1256

— 1351
— 1414
— 1473

— 1566
— 1626

— 1716

— 1805
— 1866
— 1926
— 1962

— 19:2202

— 20:2322
— 21:2402

CORONAL PLANE (90°)

HORIZONTAL PLANE (0°)   PARIETAL LOBE

PARACENTRAL LOBULE

TELENCEPHALON
CEREBRAL HEMISPHERE

OCCIPITAL LOBE

INSULA

FRONTAL LOBE

TEMPORAL LOBE

MIDBRAIN
PONS

CEREBELLUM

Olfactory bulb
Optic nerve
DIENCEPHALON (HYPOTHALAMUS)

MEDULLA

Trigeminal nerve

SPINAL CORD

Figure 8. The lateral view of the brain and upper cervical spinal cord from a specimen with a crown-rump length of 53 mm (modified from Figure 46, Table VIII, Hochstetter, 1919) serves to show the approximate locations and cutting angles of the illustrated sections of C1500 in the following pages. The small inset identifies the major structural features. The line beneath the cerebellum is the cut edge of the medullary velum.

14

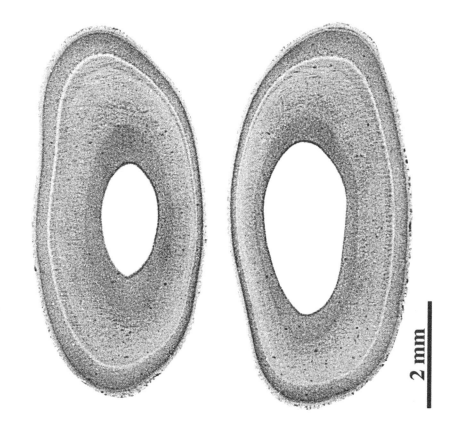

2 mm

PLATE 1A
CR 57 mm, GW 11.9, C1500
Horizontal
Section 108

PLATE 1B

FONT KEY:
VENTRICULAR DIVISIONS – CAPITALS
Germinal zone - Helvetica bold
Transient structure - Times bold italic
Permanent structure - Times Roman or **Bold**

Arrows indicate the presumed *direction of neuron migration* from neuroepithelial sources.

NEP - Neuroepithelium
SVZ - Subventricular zone

LAYERS OF THE CORTICAL
STRATIFIED TRANSITIONAL FIELD (STF)

STF1 Superficial fibrous layer with an early developmental stage (*t1*) when many cells are migrating through it, followed by a late stage (*t2*) with sparse cells. Endures as the subcortical white matter.

STF5 Deep cellular layer that is prominent during the first trimester, the first sojourn zone to appear outside the germinal matrix.

Cortical (paracentral) NEP and SVZ

FUTURE PARACENTRAL LOBULE

Layer I (contains *channel 1*)
Cortical plate
Subplate (layer VII)
Channel 2
STF1 t1
STF5
— Paracentral *STF*

Interhemispheric fissure

TELENCEPHALIC
SUPERV ENTRICLE
(FUTURE LATERAL VENTRICLE)
MID–DORSAL POOL

Cortical (paracentral) NEP and SVZ

Channel 1 is a system of glial channels in layer I; Cajal-Retzius neurons settle among them. *Channel 2* is a system of glial channels beneath the cortical plate that appears in rats on E17; subplate neurons settle among them. (Bayer and Altman, 1991). The prominent cell-sparse band breneath the cortical plate in this spcimen is postulated to be the second set of glial channels.

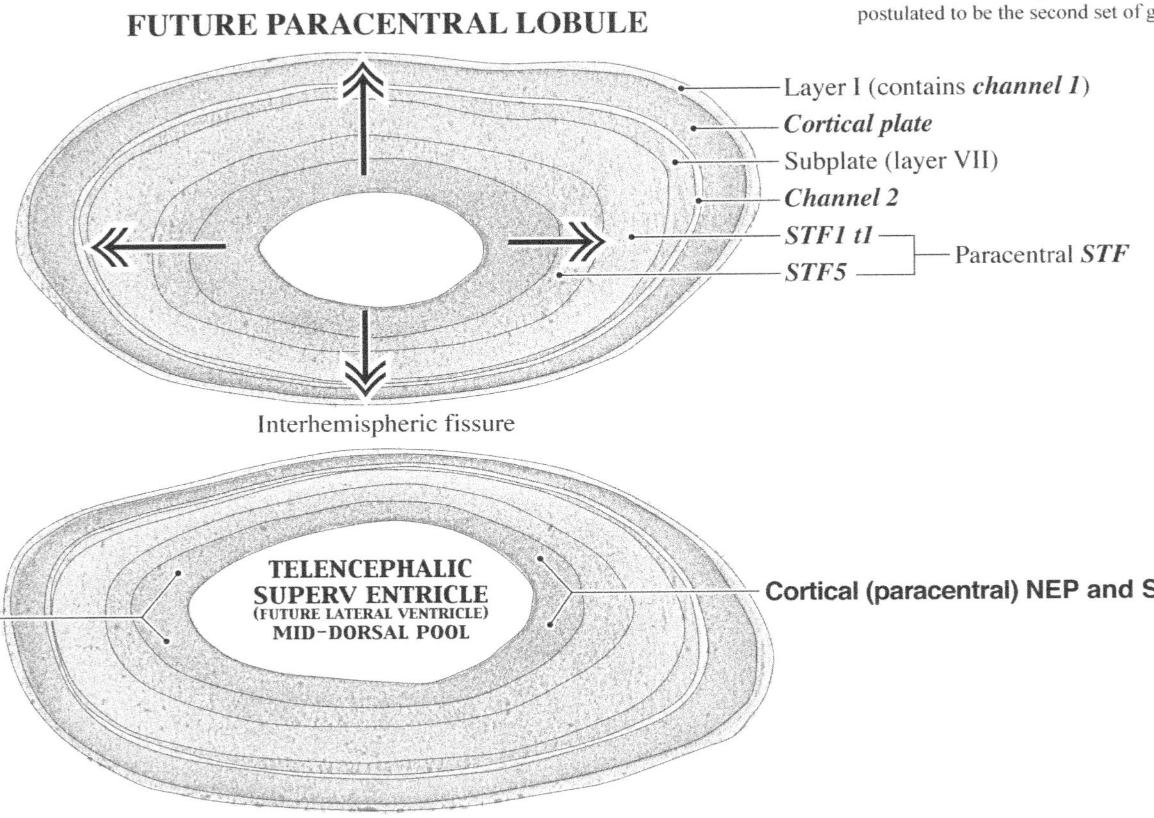

15

16

PLATE 2A
CR 57 mm, GW 11.9, C1500
Horizontal
Section 536

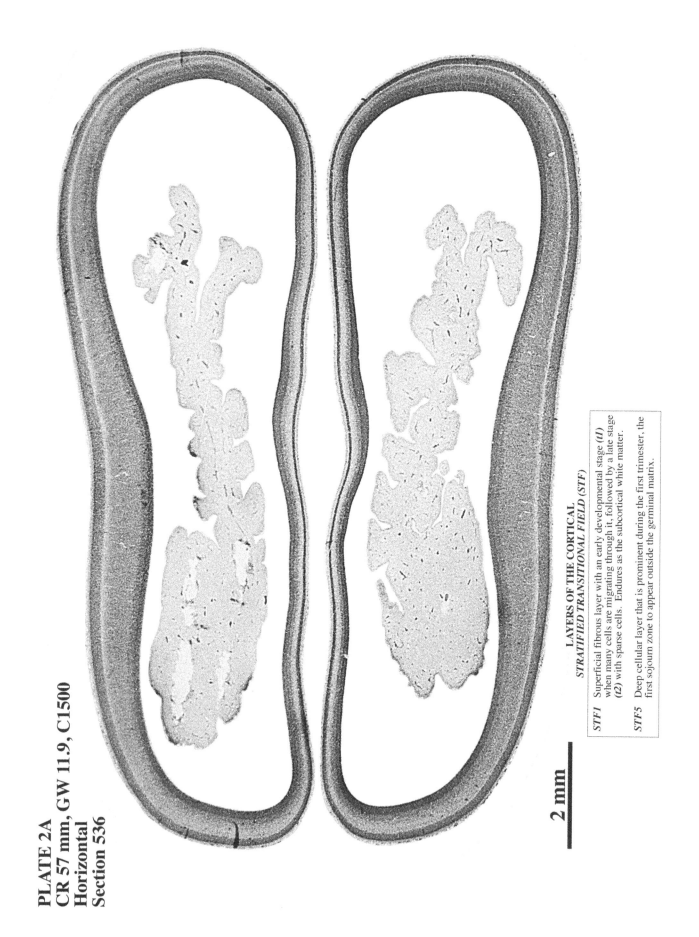

LAYERS OF THE CORTICAL
STRATIFIED TRANSITIONAL FIELD (STF)

STF1   Superficial fibrous layer with an early developmental stage (*t1*) when many cells are migrating through it, followed by a late stage (*t2*) with sparse cells. Endures as the subcortical white matter.

STF5   Deep cellular layer that is prominent during the first trimester, the first sojourn zone to appear outside the germinal matrix.

2 mm

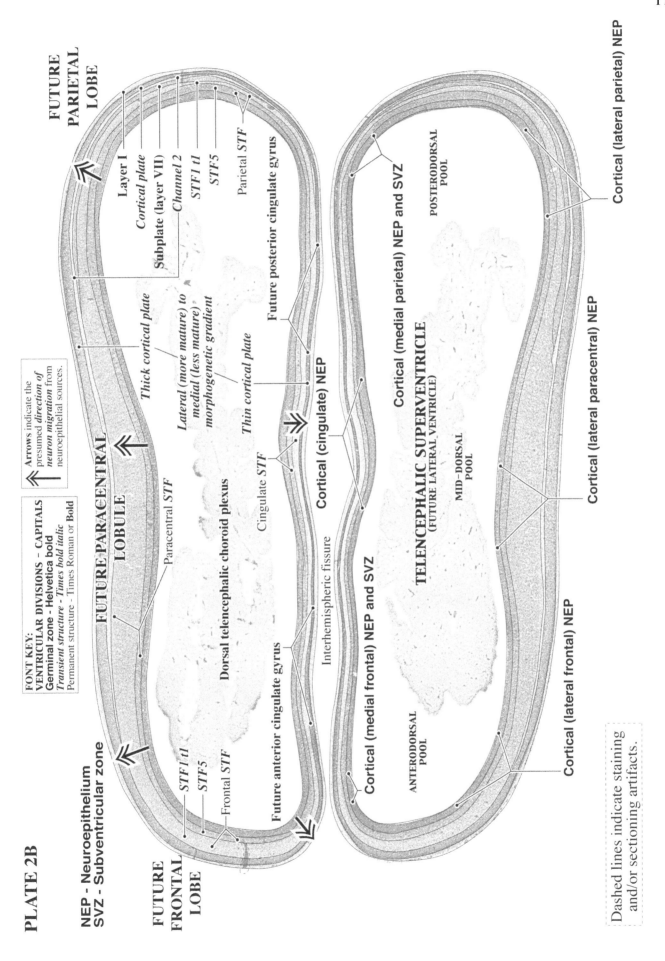

PLATE 2B

NEP - Neuroepithelium
SVZ - Subventricular zone

FONT KEY:
VENTRICULAR DIVISIONS - CAPITALS
Germinal zone - Helvetica bold
Transient structure - Times bold italic
Permanent structure - Times Roman or Bold

Arrows indicate the
presumed *direction of
neuron migration* from
neuroepithelial sources.

FUTURE
PARIETAL
LOBE

FUTURE PARACENTRAL
LOBULE

FUTURE
FRONTAL
LOBE

Layer I
Cortical plate
Subplate (layer VII)
Channel 2
STF1 t1
STF5

Parietal STF

Future posterior cingulate gyrus

Thick cortical plate

*Lateral (more mature) to
medial (less mature)
morphogenetic gradient*

Thin cortical plate

Cingulate STF

Cortical (cingulate) NEP

Paracentral STF

Dorsal telencephalic choroid plexus

Interhemispheric fissure

STF1 t1
STF5
Frontal STF

Future anterior cingulate gyrus

Cortical (medial frontal) NEP and SVZ

ANTERODORSAL
POOL

Cortical (lateral frontal) NEP

Cortical (medial parietal) NEP and SVZ

POSTERODORSAL
POOL

Cortical (lateral parietal) NEP

Cortical (medial parietal) NEP

TELENCEPHALIC SUPERVENTRICLE
(FUTURE LATERAL VENTRICLE)

MID–DORSAL
POOL

Cortical (lateral paracentral) NEP

Dashed lines indicate staining
and/or sectioning artifacts.

17

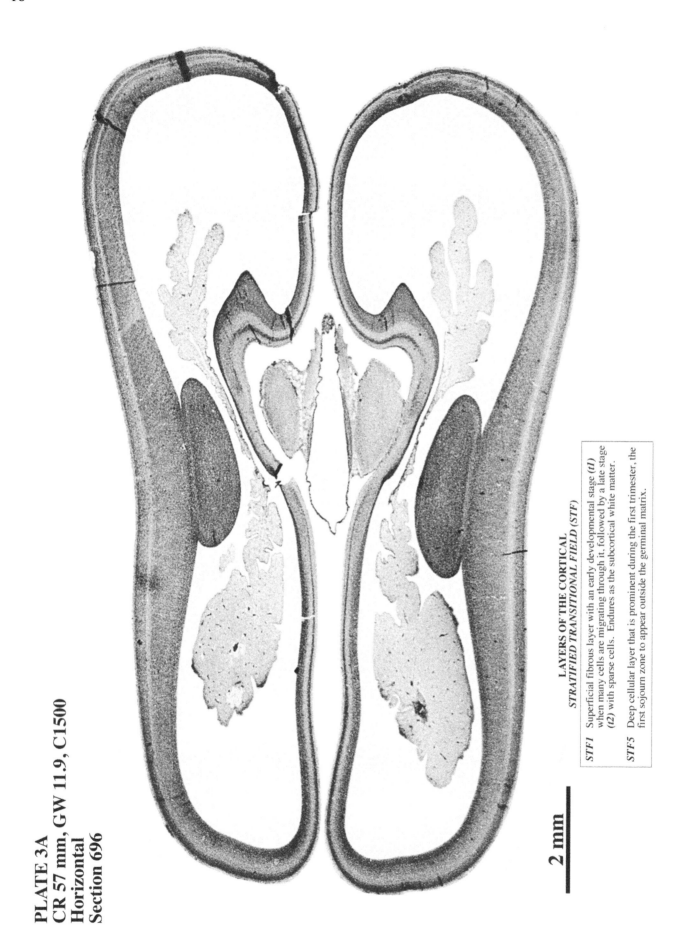

PLATE 3A
CR 57 mm, GW 11.9, C1500
Horizontal
Section 696

18

LAYERS OF THE CORTICAL
STRATIFIED TRANSITIONAL FIELD (STF)

STF1 Superficial fibrous layer with an early developmental stage (*t1*) when many cells are migrating through it, followed by a late stage (*t2*) with sparse cells. Endures as the subcortical white matter.

STF5 Deep cellular layer that is prominent during the first trimester, the first sojourn zone to appear outside the germinal matrix.

2 mm

PLATE 3B

ABBREVIATIONS:
GEP - Glioepithelium
NEP - Neuroepithelium
SVZ - Subventricular zone

FONT KEY:
VENTRICULAR DIVISIONS - CAPITALS
Germinal zone - Helvetica bold
Transient structure - Times bold italic
Permanent structure - Times Roman or Bold

⟪ Arrows indicate the presumed *direction of neuron migration* from neuroepithelial sources.

Note the lateral (more mature) to medial (less mature) morphogenetic gradient in the cerebral cortex.

Dashed lines indicate staining and/or sectioning artifacts.

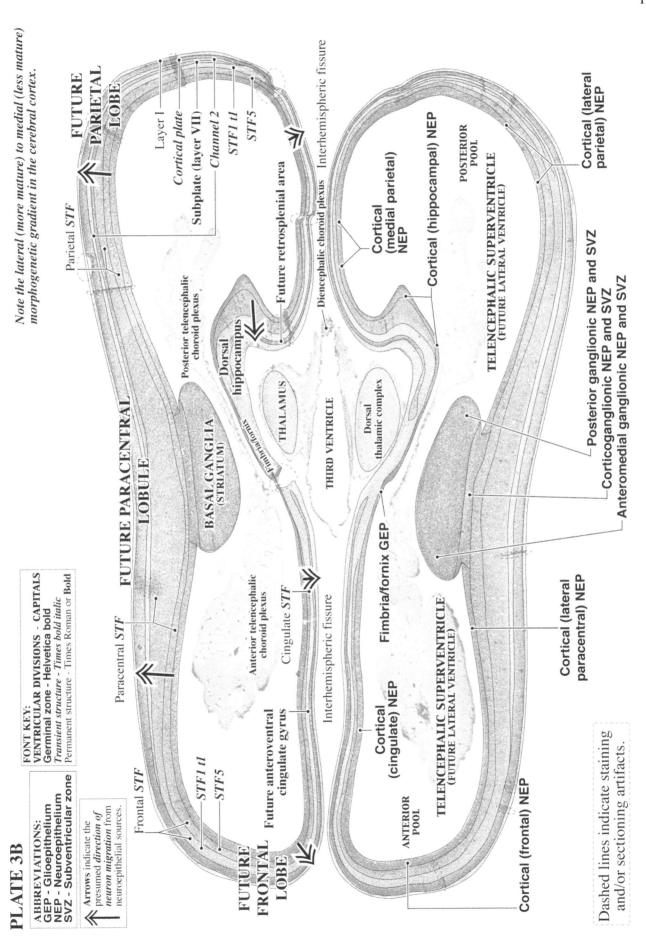

FUTURE PARIETAL LOBE

Layer I
Cortical plate
Subplate (layer VII)
Channel 2
STF1 t1
STF5

Parietal *STF*

Posterior telencephalic choroid plexus

Future retrosplenial area

Dorsal hippocampus

Interhemispheric fissure

Diencephalic choroid plexus

Cortical (medial parietal) NEP

Cortical (hippocampal) NEP

POSTERIOR POOL

TELENCEPHALIC SUPERVENTRICLE
(FUTURE LATERAL VENTRICLE)

Cortical (lateral parietal) NEP

FUTURE PARACENTRAL LOBULE

BASAL GANGLIA (STRIATUM)

Fimbria/fornix

THALAMUS

THIRD VENTRICLE

Dorsal thalamic complex

Paracentral *STF*

Posterior ganglionic NEP and SVZ
Corticoganglionic NEP and SVZ
Anteromedial ganglionic NEP and SVZ

Anterior telencephalic choroid plexus

Cingulate *STF*

Interhemispheric fissure

Future anteroventral cingulate gyrus

Cortical (lateral paracentral) NEP

Frontal *STF*

STF1 t1
STF5

Cortical (cingulate) NEP

Fimbria/fornix GEP

TELENCEPHALIC SUPERVENTRICLE
(FUTURE LATERAL VENTRICLE)

ANTERIOR POOL

FUTURE FRONTAL LOBE

Cortical (frontal) NEP

PLATE 4A
CR 57 mm, GW 11.9, C1500
Horizontal
Section 800

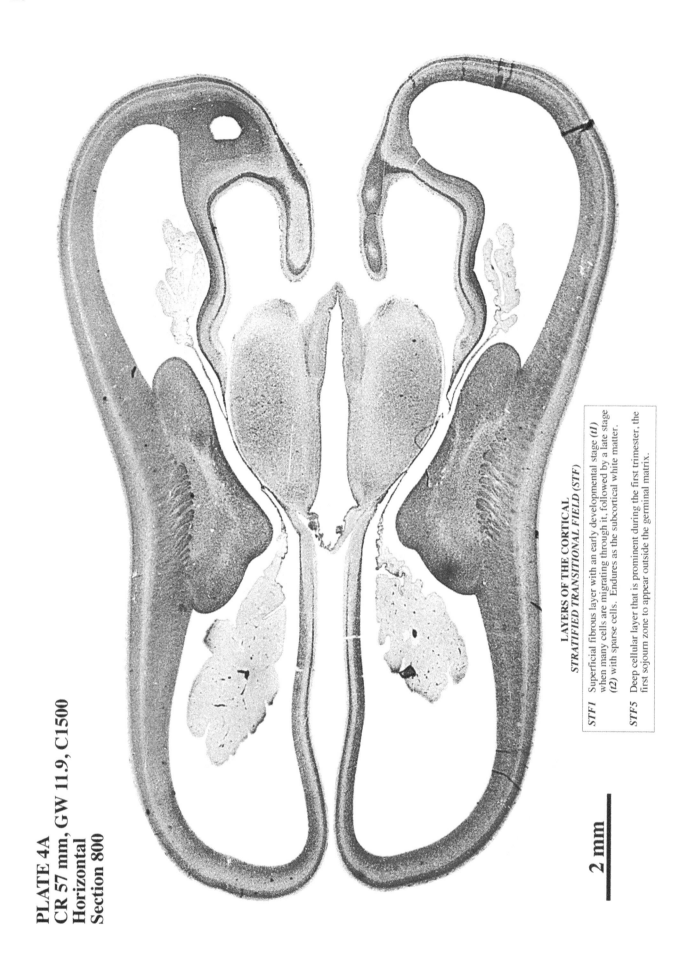

LAYERS OF THE CORTICAL
STRATIFIED TRANSITIONAL FIELD (STF)

STF1 Superficial fibrous layer with an early developmental stage (*t1*) when many cells are migrating through it, followed by a late stage (*t2*) with sparse cells. Endures as the subcortical white matter.

STF5 Deep cellular layer that is prominent during the first trimester, the first sojourn zone to appear outside the germinal matrix.

2 mm

21

PLATE 4B

Note the lateral (more mature) to medial (less mature) morphogenetic gradient in the cerebral cortex.

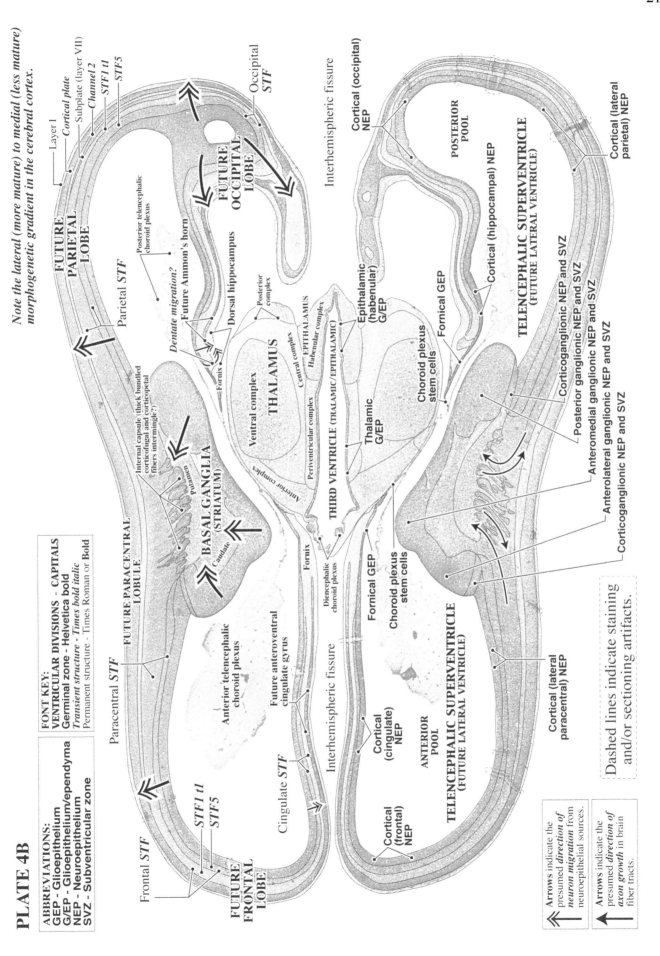

Arrows indicate the presumed *direction of neuron migration* from neuroepithelial sources.

Arrows indicate the presumed *direction of axon growth* in brain fiber tracts.

Dashed lines indicate staining and/or sectioning artifacts.

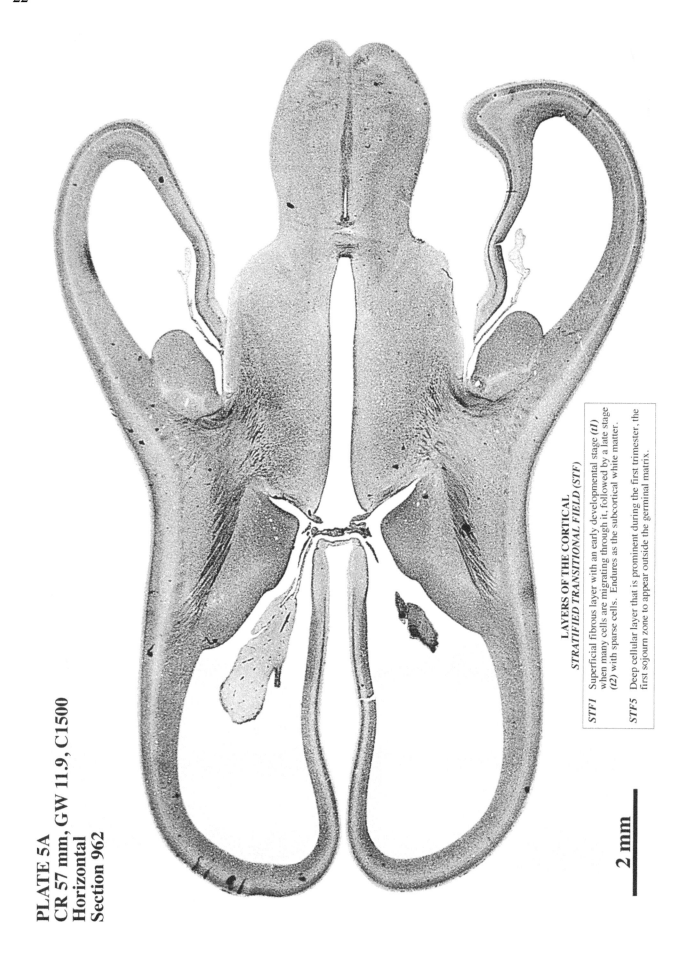

22

PLATE 5A
CR 57 mm, GW 11.9, C1500
Horizontal
Section 962

LAYERS OF THE CORTICAL
STRATIFIED TRANSITIONAL FIELD (STF)

STF1 Superficial fibrous layer with an early developmental stage (*t1*) when many cells are migrating through it, followed by a late stage (*t2*) with sparse cells. Endures as the subcortical white matter.

STF5 Deep cellular layer that is prominent during the first trimester, the first sojourn zone to appear outside the germinal matrix.

2 mm

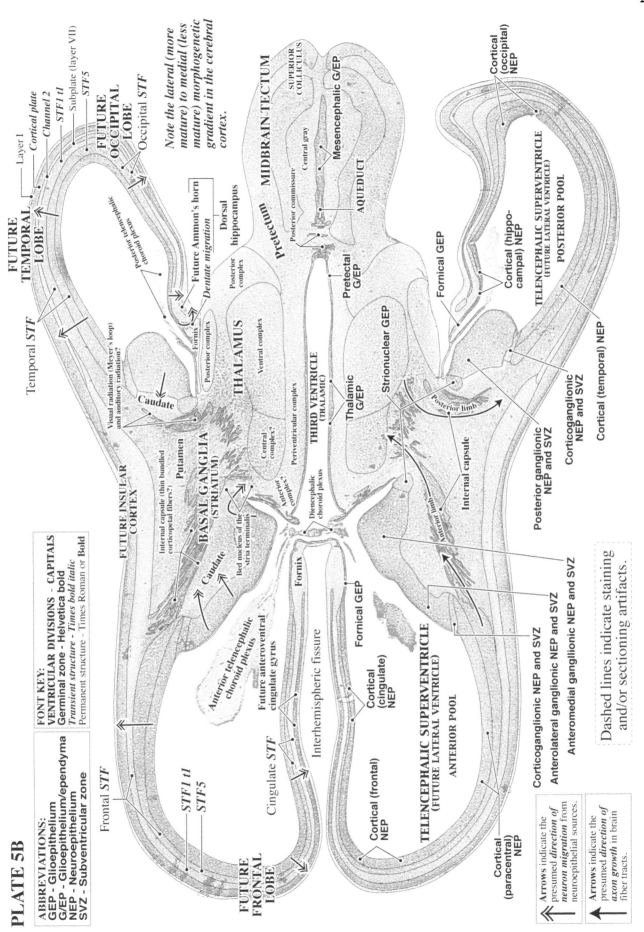

PLATE 5B

23

ABBREVIATIONS:
GEP - Glioepithelium
G/EP - Glioepithelium/ependyma
NEP - Neuroepithelium
SVZ - Subventricular zone

FONT KEY:
VENTRICULAR DIVISIONS - CAPITALS
Germinal zone - Helvetica bold
Transient structure - Times bold italic
Permanent structure - Times Roman or **Bold**

Note the lateral (more mature) to medial (less mature) morphogenetic gradient in the cerebral cortex.

Layer 1
Cortical plate
Channel 2
STF1 t1
Subplate (layer VII)
STF5

FUTURE TEMPORAL LOBE

FUTURE OCCIPITAL LOBE

Occipital *STF*

Temporal *STF*

Posterior telencephalic choroid plexus

Future Ammon's horn

Dorsal hippocampus

MIDBRAIN TECTUM

Dentate migration

SUPERIOR COLLICULUS

Fornix

Posterior complex

Pretectum

Mesencephalic G/EP

Posterior commissure

Central gray

THALAMUS

Ventral complex

AQUEDUCT

Posterior complex

Periventricular complex

Pretectal G/EP

Visual radiation (Meyer's loop) and auditory radiation?

Caudate

Central complex?

Thalamic G/EP

Fornical GEP

FUTURE INSULAR CORTEX

Internal capsule (thin bundled corticopetal fibers?)

Putamen

Strionuclear GEP

THIRD VENTRICLE (THALAMIC)

Cortical (hippo-campal) NEP

TELENCEPHALIC SUPERVENTRICLE (FUTURE LATERAL VENTRICLE) POSTERIOR POOL

BASAL GANGLIA (STRIATUM)

Central complex?

Anterior complex?

Periventricular complex

Posterior limb

Caudate

Bed nucleus of the stria terminalis

Diencephalic choroid plexus

Internal capsule

Posterior ganglionic NEP and SVZ

Corticoganglionic NEP and SVZ

Cortical (temporal) NEP

Frontal *STF*

Fornix

Anterior telencephalic choroid plexus

Future anteroventral cingulate gyrus

Interhemispheric fissure

Fornical GEP

Anterior limb

TELENCEPHALIC SUPERVENTRICLE (FUTURE LATERAL VENTRICLE) ANTERIOR POOL

Cortical (cingulate) NEP

Corticoganglionic NEP and SVZ
Anterolateral ganglionic NEP and SVZ
Anteromedial ganglionic NEP and SVZ

STF1 t1
STF5

Cingulate *STF*

Cortical (frontal) NEP

Cortical (paracentral) NEP

FUTURE FRONTAL LOBE

Dashed lines indicate staining and/or sectioning artifacts.

≪ **Arrows** indicate the presumed *direction of neuron migration* from neuroepithelial sources.

◄ **Arrows** indicate the presumed *direction of axon growth* in brain fiber tracts.

24

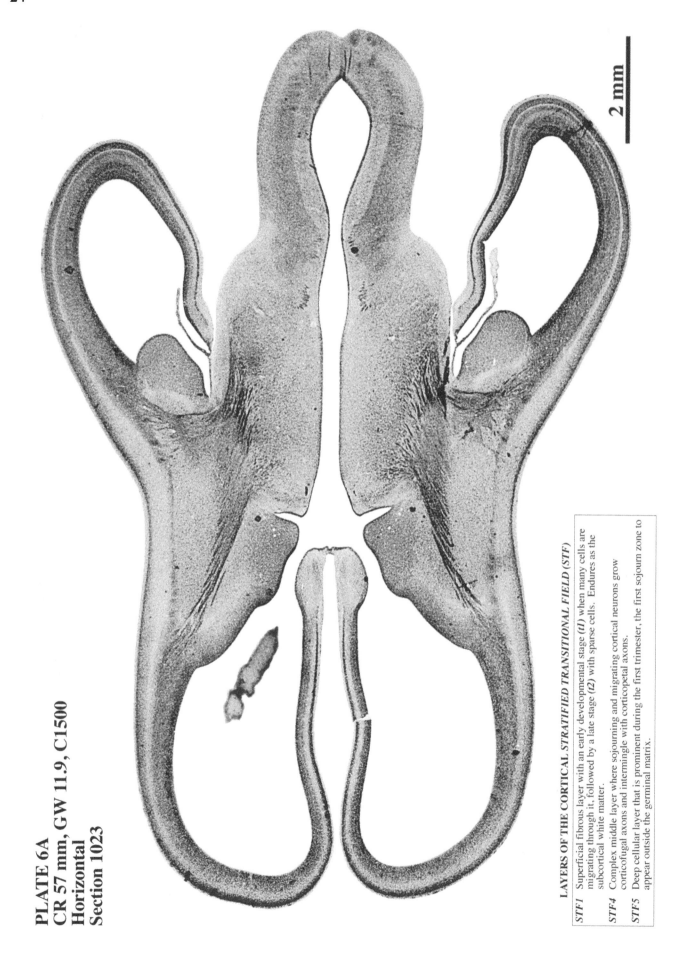

PLATE 6A
CR 57 mm, GW 11.9, C1500
Horizontal
Section 1023

2 mm

LAYERS OF THE CORTICAL *STRATIFIED TRANSITIONAL FIELD (STF)*

STF1 Superficial fibrous layer with an early developmental stage (*t1*) when many cells are migrating through it, followed by a late stage (*t2*) with sparse cells. Endures as the subcortical white matter.

STF4 Complex middle layer where sojourning and migrating cortical neurons grow corticofugal axons and intermingle with corticopetal axons.

STF5 Deep cellular layer that is prominent during the first trimester, the first sojourn zone to appear outside the germinal matrix.

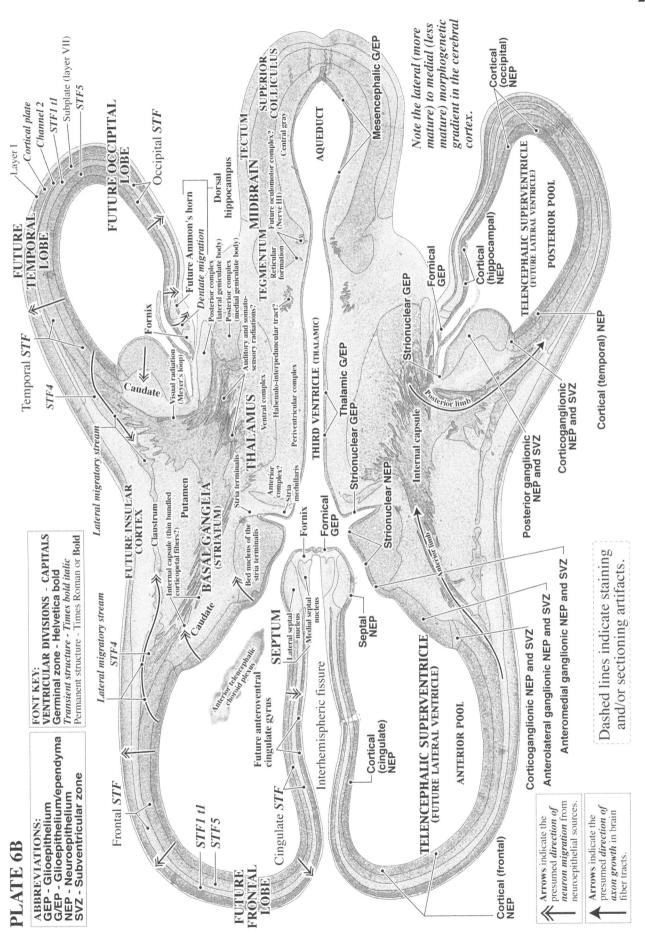

PLATE 6B

25

PLATE 7A
CR 57 mm, GW 11.9, C1500
Horizontal
Section 1143

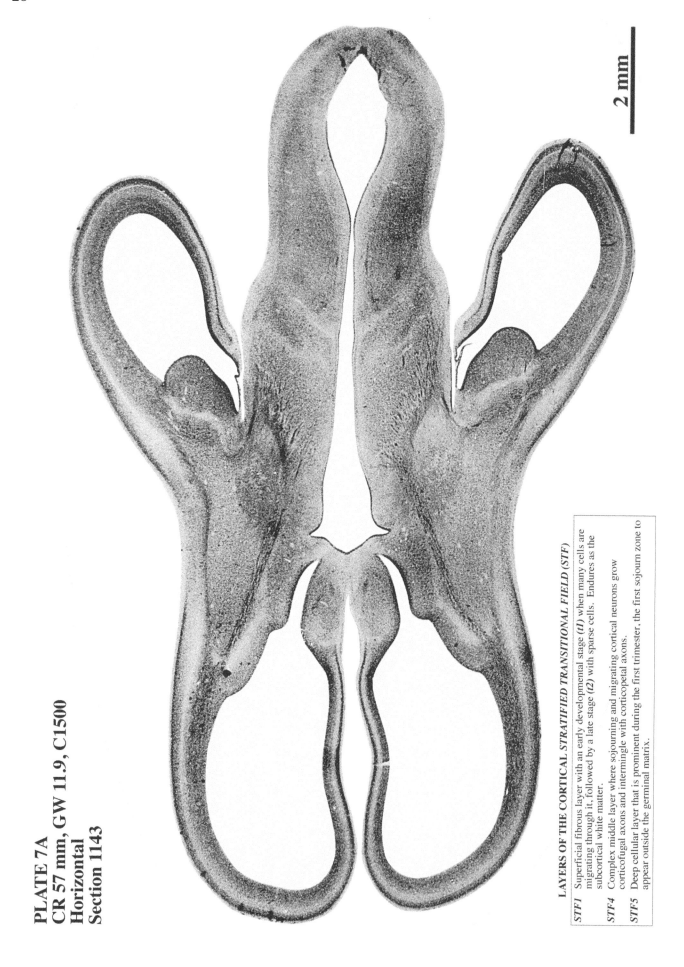

LAYERS OF THE CORTICAL *STRATIFIED TRANSITIONAL FIELD (STF)*

STF1 Superficial fibrous layer with an early developmental stage (*t1*) when many cells are migrating through it, followed by a late stage (*t2*) with sparse cells. Endures as the subcortical white matter.

STF4 Complex middle layer where sojourning and migrating cortical neurons grow corticofugal axons and intermingle with corticopetal axons.

STF5 Deep cellular layer that is prominent during the first trimester, the first sojourn zone to appear outside the germinal matrix.

2 mm

PLATE 7B

ABBREVIATIONS:
GEP - Glioepithelium
G/EP - Glioepithelium/ependyma
NEP - Neuroepithelium
SVZ - Subventricular zone

FONT KEY:
VENTRICULAR DIVISIONS - CAPITALS
Germinal zone - Helvetica bold
Transient structure - Times bold italic
Permanent structure - Times Roman or Bold

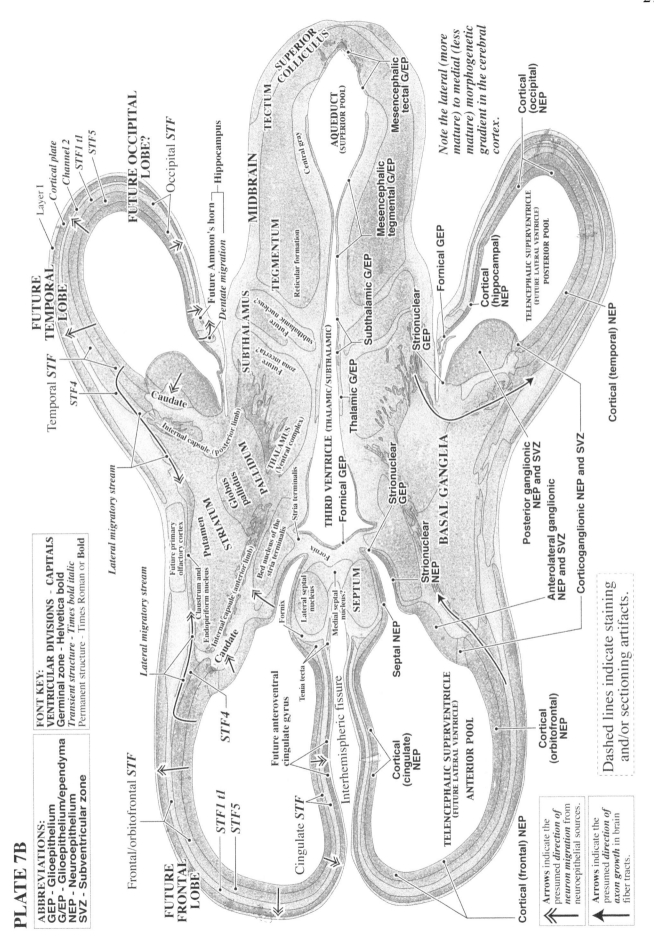

Note the lateral (more mature) to medial (less mature) morphogenetic gradient in the cerebral cortex.

Dashed lines indicate staining and/or sectioning artifacts.

Arrows indicate the presumed *direction of neuron migration* from neuroepithelial sources.

Arrows indicate the presumed *direction of axon growth* in brain fiber tracts.

Layer I
Cortical plate
Channel 2
STF1 t1
STF5

FUTURE OCCIPITAL LOBE?
Occipital *STF*
Hippocampus
Future Ammon's horn
Dentate migration

FUTURE TEMPORAL LOBE
Temporal *STF*
STF4

Lateral migratory stream
Caudate
Internal capsule (Posterior limb)

Frontal/orbitofrontal STF

FUTURE FRONTAL LOBE
STF1 t1
STF5
Cingulate *STF*
STF4
Lateral migratory stream

Future primary olfactory cortex
Claustrum and Endopiriform nucleus
Internal capsule (anterior limb)
Caudate

STRIATUM
Putamen
PALLIDUM
Globus pallidus
Bed nucleus of the stria terminalis
Stria terminalis

Future anteroventral cingulate gyrus
Tenia tecta
Fornix

Interhemispheric fissure
Cortical (cingulate) NEP

SUPERIOR COLLICULUS
TECTUM
MIDBRAIN
Central gray
AQUEDUCT (SUPERIOR POOL)
Mesencephalic tectal G/EP

TEGMENTUM
Reticular formation
Mesencephalic tegmental G/EP

SUBTHALAMUS
Future subthalamic nucleus?
Future zona incerta?

THALAMUS (Ventral complex)

Fornix
Lateral septal nucleus
Medial septal nucleus?
SEPTUM
Septal NEP

THIRD VENTRICLE (THALAMIC/SUBTHALAMIC)
Fornical GEP
Thalamic G/EP
Subthalamic G/EP
Strionuclear GEP
Fornical GEP
Strionuclear GEP
Strionuclear NEP

BASAL GANGLIA
Strionuclear GEP
Posterior ganglionic NEP and SVZ
Anterolateral ganglionic NEP and SVZ
Corticoganglionic NEP and SVZ

Cortical (hippocampal) NEP
TELENCEPHALIC SUPERVENTRICLE (FUTURE LATERAL VENTRICLE) POSTERIOR POOL

Cortical (occipital) NEP
Cortical (temporal) NEP

Cortical (orbitofrontal) NEP
TELENCEPHALIC SUPERVENTRICLE (FUTURE LATERAL VENTRICLE) ANTERIOR POOL

Cortical (frontal) NEP

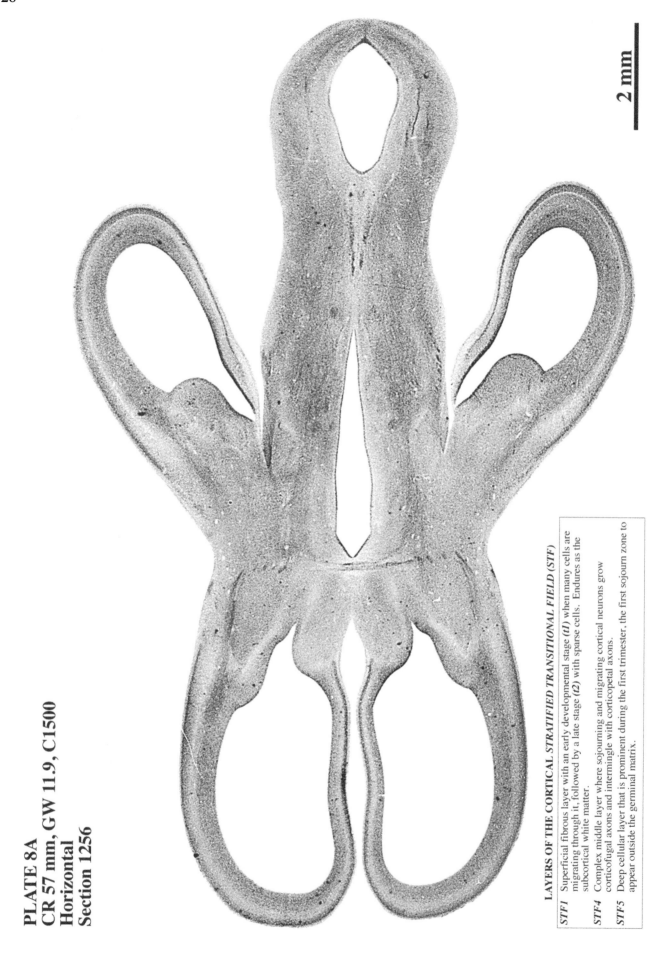

28

PLATE 8A
CR 57 mm, GW 11.9, C1500
Horizontal
Section 1256

2 mm

LAYERS OF THE CORTICAL *STRATIFIED TRANSITIONAL FIELD* (STF)

STF1   Superficial fibrous layer with an early developmental stage (*t1*) when many cells are migrating through it, followed by a late stage (*t2*) with sparse cells. Endures as the subcortical white matter.

STF4   Complex middle layer where sojourning and migrating cortical neurons grow corticofugal axons and intermingle with corticopetal axons.

STF5   Deep cellular layer that is prominent during the first trimester, the first sojourn zone to appear outside the germinal matrix.

PLATE 8B

ABBREVIATIONS:
GEP - Glioepithelium
G/EP - Glioepithelium/ependyma
NEP - Neuroepithelium
SVZ - Subventricular zone

FONT KEY:
VENTRICULAR DIVISIONS - CAPITALS
Germinal zone - Helvetica bold
Transient structure - Times bold italic
Permanent structure - Times Roman or Bold

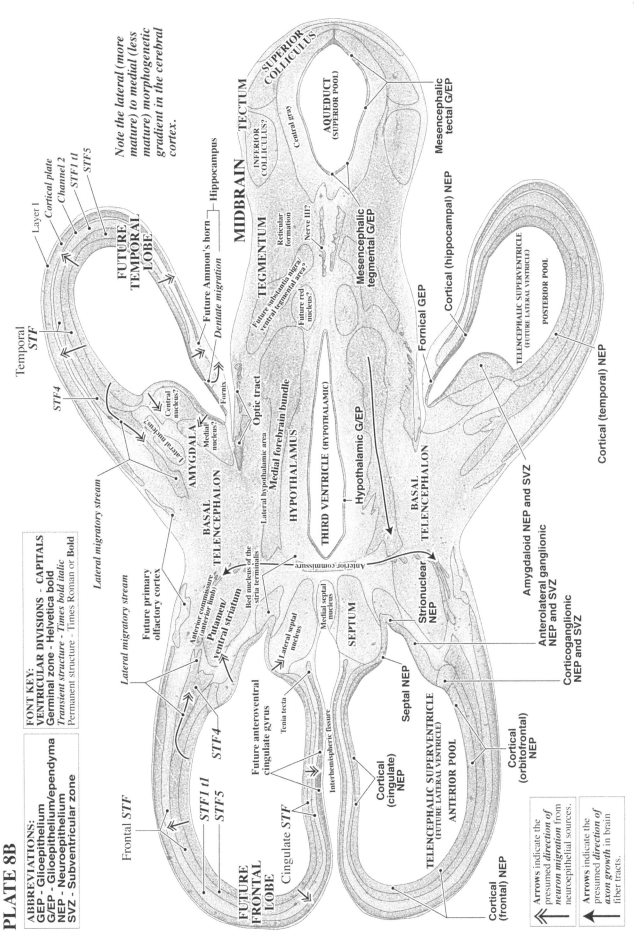

Note the lateral (more mature) to medial (less mature) morphogenetic gradient in the cerebral cortex.

Arrows indicate the presumed *direction of neuron migration* from neuroepithelial sources.

Arrows indicate the presumed *direction of axon growth* in brain fiber tracts.

30

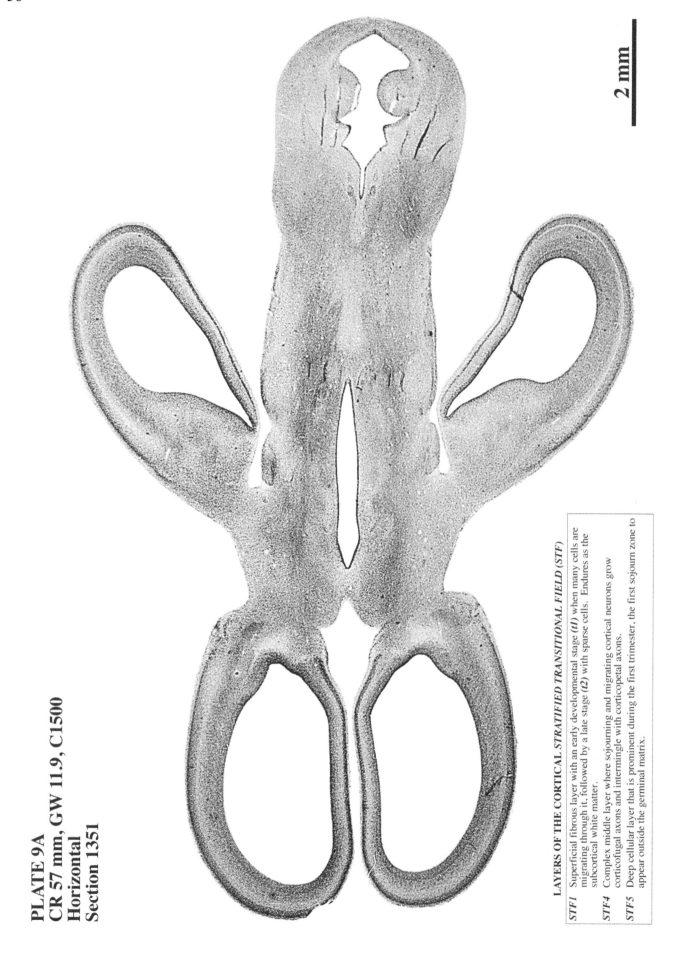

PLATE 9A
CR 57 mm, GW 11.9, C1500
Horizontal
Section 1351

2 mm

LAYERS OF THE CORTICAL *STRATIFIED TRANSITIONAL FIELD* (STF)

STF1 Superficial fibrous layer with an early developmental stage (*t1*) when many cells are migrating through it, followed by a late stage (*t2*) with sparse cells. Endures as the subcortical white matter.

STF4 Complex middle layer where sojourning and migrating cortical neurons grow corticofugal axons and intermingle with corticopetal axons.

STF5 Deep cellular layer that is prominent during the first trimester, the first sojourn zone to appear outside the germinal matrix.

PLATE 9B

ABBREVIATIONS:
G/EP - Glioepithelium/ependyma
NEP - Neuroepithelium
SVZ - Subventricular zone

FONT KEY:
VENTRICULAR DIVISIONS - CAPITALS
Germinal zone - **Helvetica bold**
Transient structure - Times bold italic
Permanent structure - Times Roman or **Bold**

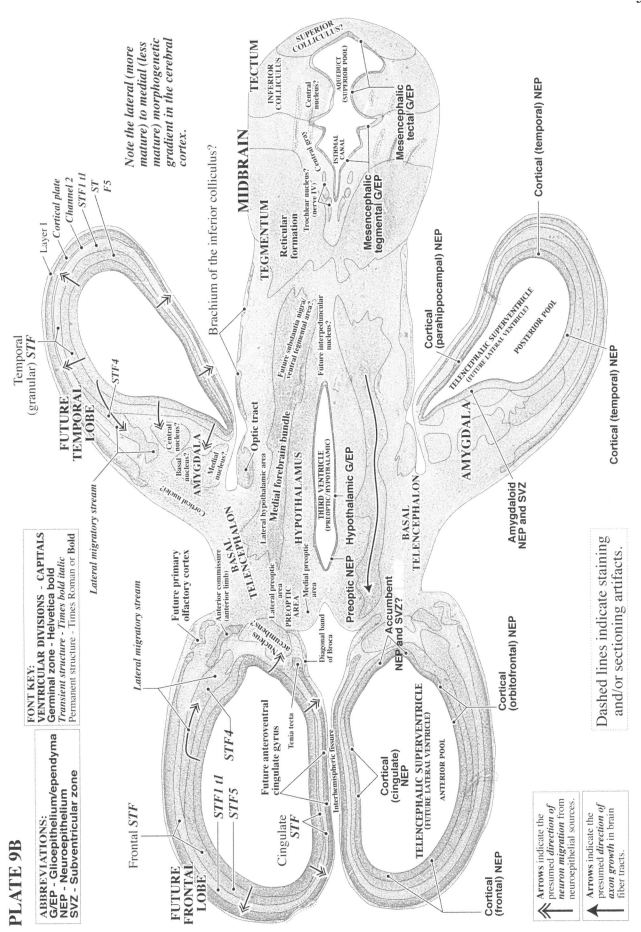

Note the lateral (more mature) to medial (less mature) morphogenetic gradient in the cerebral cortex.

Arrows indicate the presumed *direction of neuron migration* from neuroepithelial sources.

Arrows indicate the presumed *direction of axon growth* in brain fiber tracts.

Dashed lines indicate staining and/or sectioning artifacts.

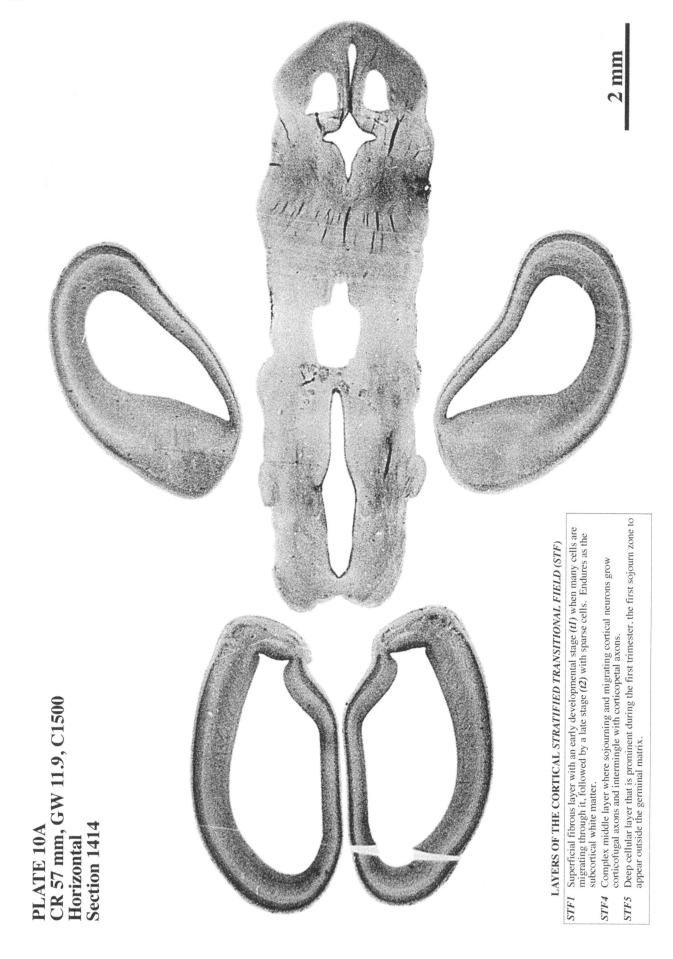

32

PLATE 10A
CR 57 mm, GW 11.9, C1500
Horizontal
Section 1414

LAYERS OF THE CORTICAL *STRATIFIED TRANSITIONAL FIELD* (STF)

STF1 Superficial fibrous layer with an early developmental stage (*t1*) when many cells are migrating through it, followed by a late stage (*t2*) with sparse cells. Endures as the subcortical white matter.

STF4 Complex middle layer where sojourning and migrating cortical neurons grow corticofugal axons and intermingle with corticopetal axons.

STF5 Deep cellular layer that is prominent during the first trimester, the first sojourn zone to appear outside the germinal matrix.

2 mm

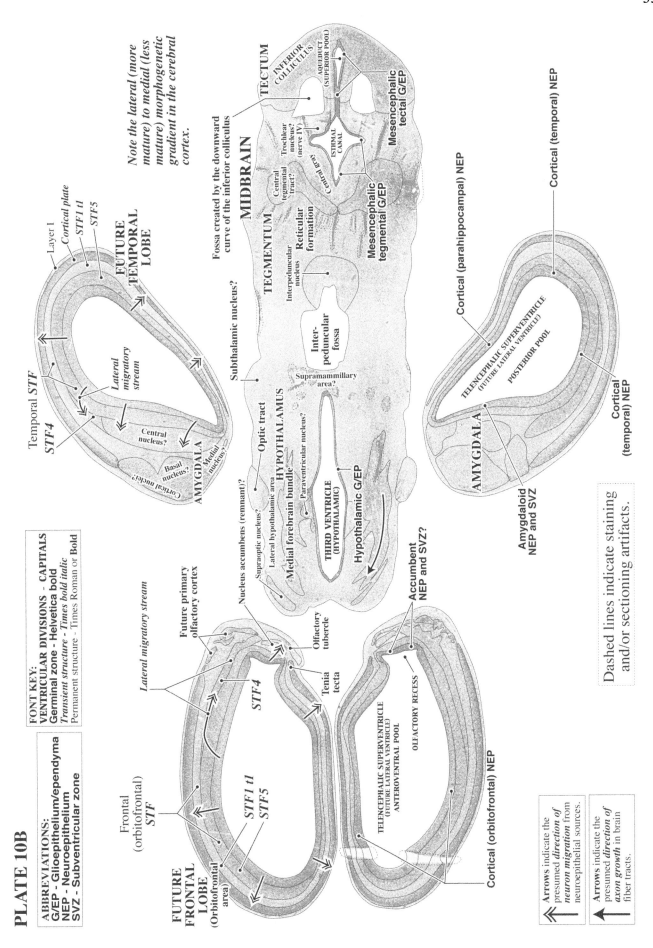

PLATE 10B

ABBREVIATIONS:
G/EP - Glioepithelium/ependyma
NEP - Neuroepithelium
SVZ - Subventricular zone

FONT KEY:
VENTRICULAR DIVISIONS - CAPITALS
Germinal zone - Helvetica bold
Transient structure - Times bold italic
Permanent structure - Times Roman or **Bold**

Note the lateral (more mature) to medial (less mature) morphogenetic gradient in the cerebral cortex.

Layer 1
Cortical plate
STF1 t1
STF5

FUTURE TEMPORAL LOBE

Temporal *STF*

STF4

Lateral migratory stream

Central nucleus?

Cortical nuclei?

Basal nucleus?

Medial nucleus?

AMYGDALA

Fossa created by the downward curve of the inferior colliculus

TECTUM

INFERIOR COLLICULUS

AQUEDUCT (SUPERIOR POOL)

Mesencephalic tectal G/EP

Trochlear nucleus? (nerve IV)

ISTHMAL CANAL

Central gray

MIDBRAIN

Central tegmental tract?

TEGMENTUM

Reticular formation

Mesencephalic tegmental G/EP

Interpeduncular nucleus

Subthalamic nucleus?

Inter-peduncular fossa

Supramammillary area?

Optic tract

HYPOTHALAMUS

Nucleus accumbens (remnant)?

Supraoptic nucleus?

Lateral hypothalamic area

Paraventricular nucleus?

Medial forebrain bundle

THIRD VENTRICLE (HYPOTHALAMIC)

Hypothalamic G/EP

Cortical (temporal) NEP

Cortical (parahippocampal) NEP

TELENCEPHALIC SUPERVENTRICLE (FUTURE LATERAL VENTRICLE) POSTERIOR POOL

AMYGDALA

Amygdaloid NEP and SVZ

Cortical (temporal) NEP

Lateral migratory stream

Future primary olfactory cortex

Olfactory tubercle

Tenia tecta

STF4

Accumbent NEP and SVZ?

Frontal (orbitofrontal) *STF*

STF1 t1
STF5

Cortical (orbitofrontal) area

FUTURE FRONTAL LOBE

TELENCEPHALIC SUPERVENTRICLE (FUTURE LATERAL VENTRICLE) ANTEROVENTRAL POOL

OLFACTORY RECESS

Cortical (orbitofrontal) NEP

Dashed lines indicate staining and/or sectioning artifacts.

Arrows indicate the presumed *direction of neuron migration* from neuroepithelial sources.

Arrows indicate the presumed *direction of axon growth* in brain fiber tracts.

33

PLATE 11A
CR 57 mm, GW 11.9, C1500
Horizontal
Section 1473

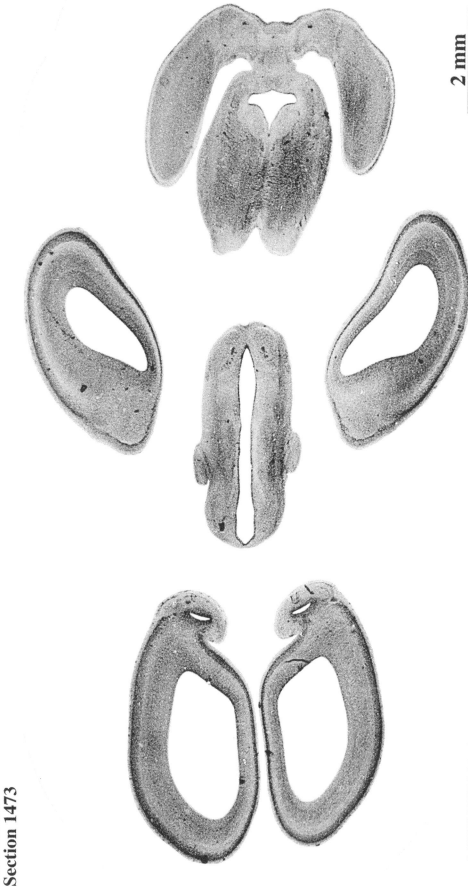

2 mm

LAYERS OF THE CORTICAL *STRATIFIED TRANSITIONAL FIELD (STF)*

STF1 Superficial fibrous layer with an early developmental stage (*t1*) when many cells are migrating through it, followed by a late stage (*t2*) with sparse cells. Endures as the subcortical white matter.

STF4 Complex middle layer where sojourning and migrating cortical neurons grow corticofugal axons and intermingle with corticopetal axons.

STF5 Deep cellular layer that is prominent during the first trimester, the first sojourn zone to appear outside the germinal matrix.

PLATE 11B

ABBREVIATIONS:
G/EP - Glioepithelium/ependyma
NEP - Neuroepithelium
SVZ - Subventricular zone

FONT KEY:
VENTRICULAR DIVISIONS - CAPITALS
Germinal zone - Helvetica bold
Transient structure - Times bold italic
Permanent structure - Times Roman or **Bold**

Note the lateral (more mature) to medial (less mature) morphogenetic gradient in the cerebral cortex.

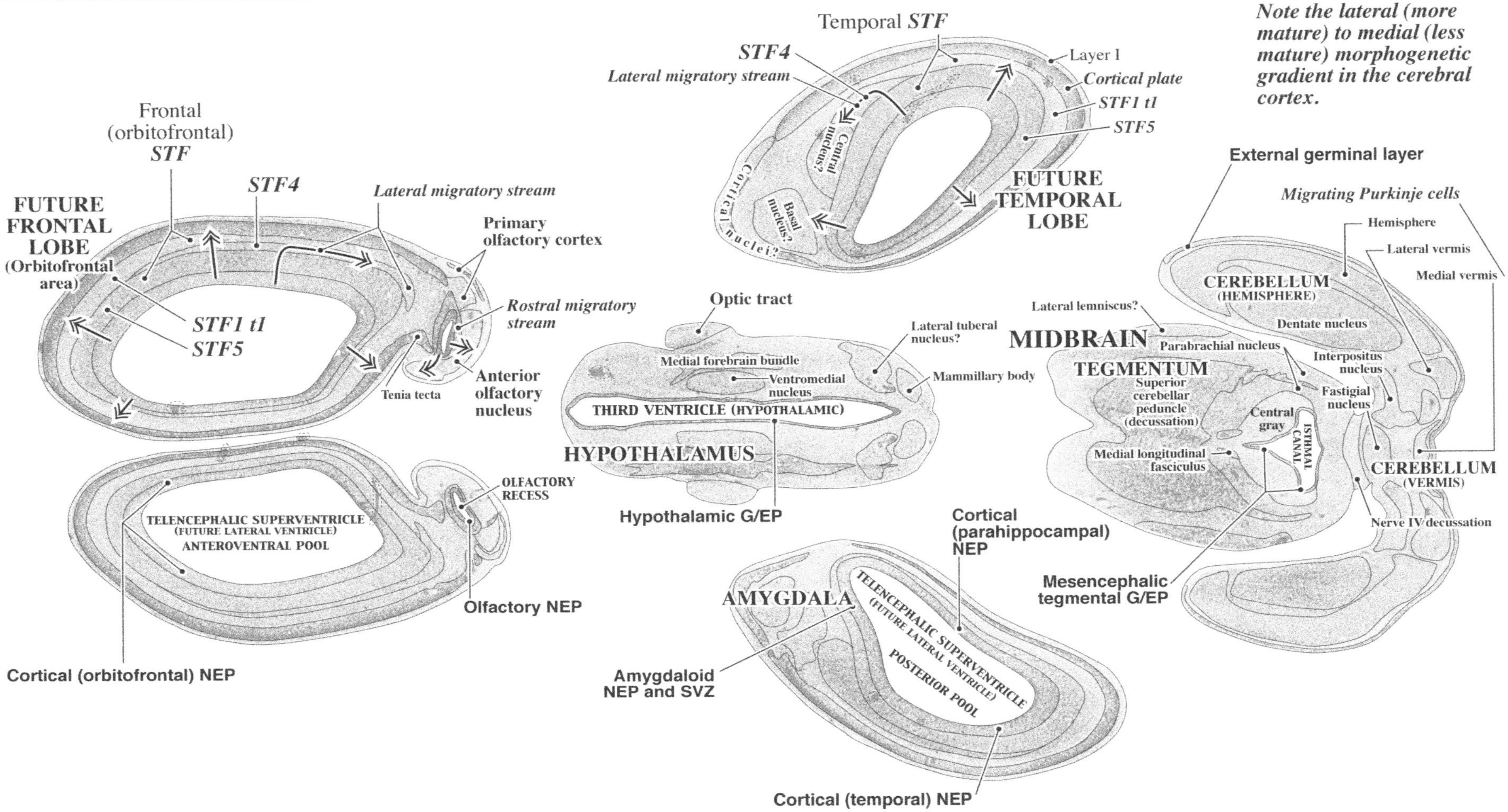

Frontal
(orbitofrontal)
STF
FUTURE
FRONTAL
LOBE
(Orbitofrontal
area)
STF4
Lateral migratory stream
Primary
olfactory cortex
Rostral migratory
stream
STF1 t1
STF5
Tenia tecta
Anterior
olfactory
nucleus
OLFACTORY
RECESS
TELENCEPHALIC SUPERVENTRICLE
(FUTURE LATERAL VENTRICLE)
ANTEROVENTRAL POOL
Olfactory NEP
Cortical (orbitofrontal) NEP
Temporal STF
STF4
Lateral migratory stream
Layer I
Cortical plate
STF1 t1
STF5
Central
nucleus?
Basal
nucleus?
Cortical nuclei?
FUTURE
TEMPORAL
LOBE
Optic tract
Medial forebrain bundle
Ventromedial
nucleus
Lateral tuberal
nucleus?
Mammillary body
THIRD VENTRICLE (HYPOTHALAMIC)
HYPOTHALAMUS
Hypothalamic G/EP
AMYGDALA
TELENCEPHALIC SUPERVENTRICLE
(FUTURE LATERAL VENTRICLE)
POSTERIOR POOL
Cortical
(parahippocampal)
NEP
Amygdaloid
NEP and SVZ
Cortical (temporal) NEP
External germinal layer
Migrating Purkinje cells
Hemisphere
Lateral vermis
Medial vermis
CEREBELLUM
(HEMISPHERE)
Dentate nucleus
Lateral lemniscus?
MIDBRAIN
Parabrachial nucleus
Interpositus
nucleus
TEGMENTUM
Superior
cerebellar
peduncle
(decussation)
Fastigial
nucleus
Central
gray
ISTHMAL
CANAL
Medial longitudinal
fasciculus
CEREBELLUM
(VERMIS)
Nerve IV decussation
Mesencephalic
tegmental G/EP

Arrows indicate the presumed ***direction of neuron migration*** from neuroepithelial sources.

Dashed lines indicate staining and/or sectioning artifacts.

36

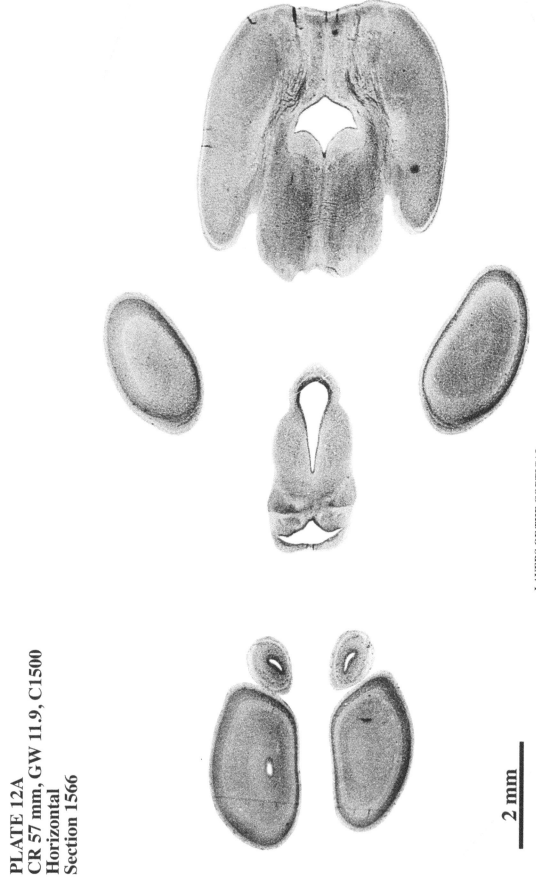

PLATE 12A
CR 57 mm, GW 11.9, C1500
Horizontal
Section 1566

2 mm

LAYERS OF THE CORTICAL
STRATIFIED TRANSITIONAL FIELD (STF)

STF1   Superficial fibrous layer with an early developmental stage (*t1*)
when many cells are migrating through it, followed by a late stage
(*t2*) with sparse cells. Endures as the subcortical white matter.

STF5   Deep cellular layer that is prominent during the first trimester, the
first sojourn zone to appear outside the germinal matrix.

PLATE 12B

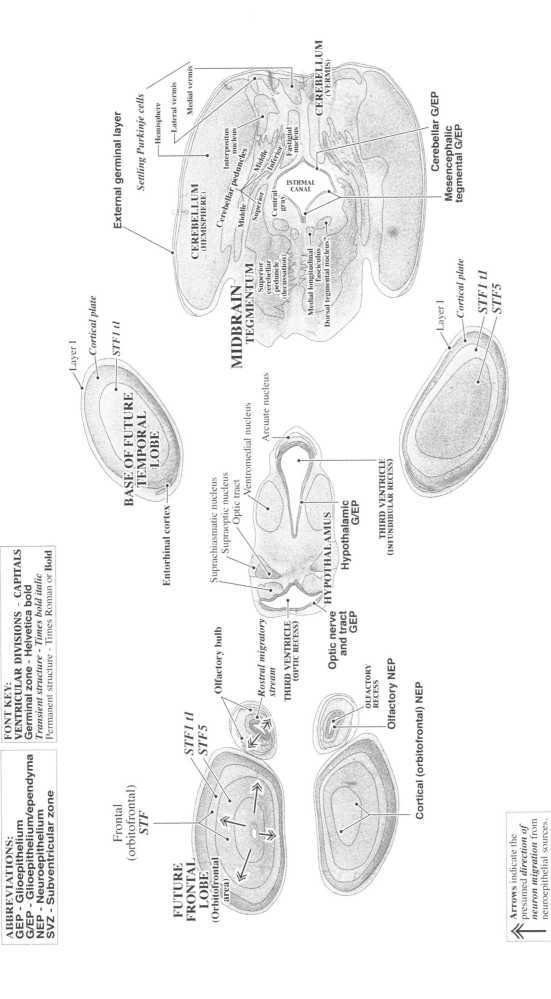

Arrows indicate the
presumed *direction of
neuron migration* from
neuroepithelial sources.

Dashed lines indicate staining
and/or sectioning artifacts.

38

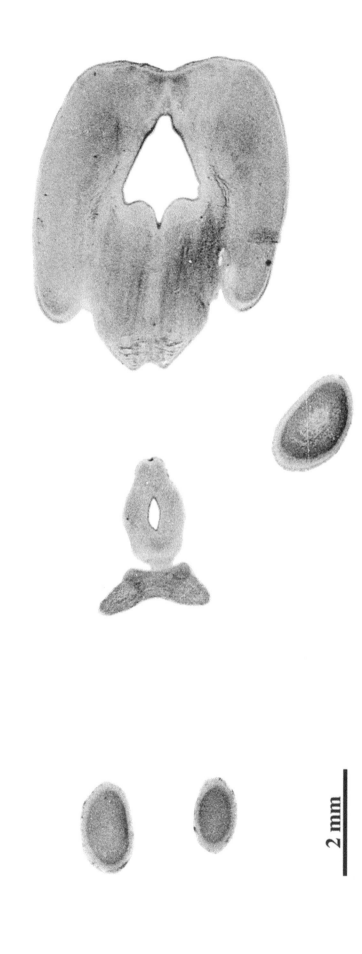

PLATE 13A
CR 57 mm, GW11.9, C1500
Horizontal
Section 1626

2 mm

PLATE 13B

FONT KEY:
VENTRICULAR DIVISIONS – CAPITALS
Germinal zone - Helvetica bold
Transient structure - Times bold italic
Permanent structure - Times Roman or **Bold**

G/EP - Glioepithelium/ependyma
NEP - Neuroepithelium

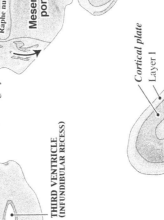

External germinal layer

Settling Purkinje cells

Hemisphere

Lateral vermis

Medial vermis

CEREBELLUM
(VERMIS)

Cerebellar NEP

CEREBELLUM
(HEMISPHERE)

Inferior cerebellar peduncle

ISTHMAL
CANAL

Central gray

Reticular formation

Raphe nuclear complex

Mesencephalic/
pontine G/EP

Olivo-cerebellar fibers?

Lateral lemniscus

MIDBRAIN/PONS
INTERFACE

Pontocerebellar
fibers

Pontine gray

Note that corticofugal fibers
are *absent* in the pontine gray.

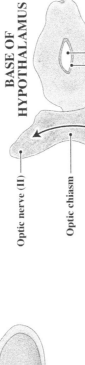

BASE OF
HYPOTHALAMUS

THIRD VENTRICLE
(INFUNDIBULAR RECESS)

Hypothalamic G/EP

Optic nerve (II)

Optic chiasm

Cortical plate

Layer I

BASE OF FUTURE
TEMPORAL LOBE

BASE OF FUTURE
FRONTAL LOBE
(Orbitofrontal area)

Layer I

Cortical plate

Arrows indicate the
presumed *direction of*
axon growth in brain
fiber tracts.

Dashed lines indicate staining
and/or sectioning artifacts.

40

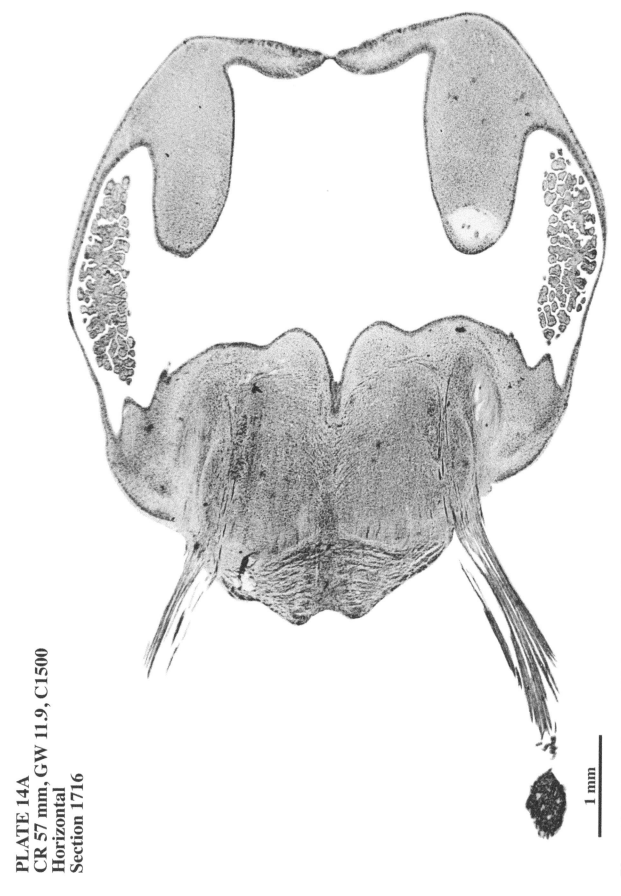

PLATE 14A
CR 57 mm, GW 11.9, C1500
Horizontal
Section 1716

Plates 14 to 21 are shown at higher magnification.

1 mm

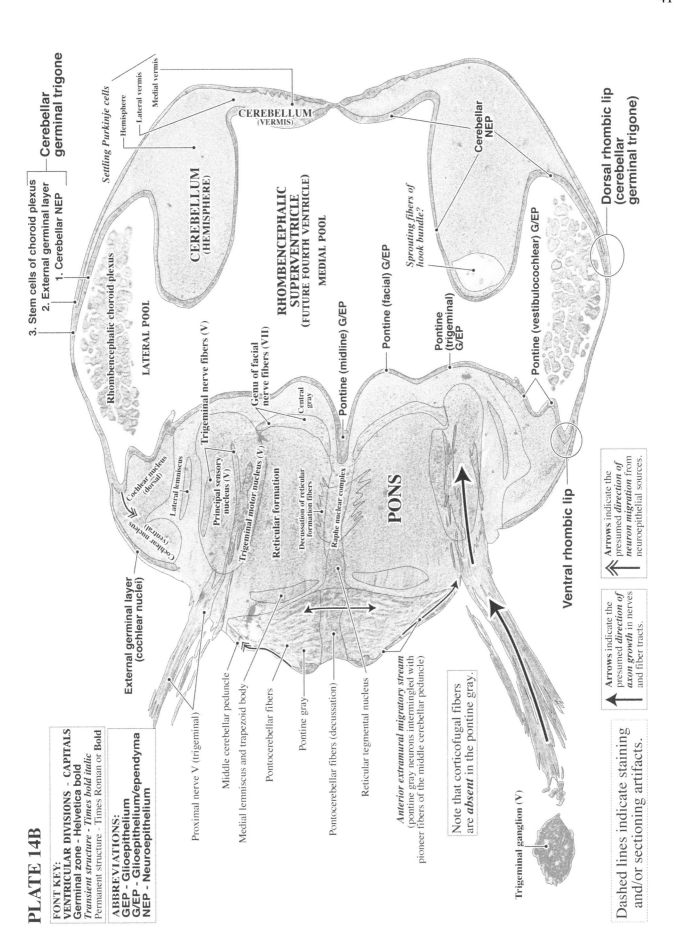

PLATE 14B

41

FONT KEY:
VENTRICULAR DIVISIONS - CAPITALS
Germinal zone - Helvetica bold
Transient structure - Times bold italic
Permanent structure - Times Roman or **Bold**

ABBREVIATIONS:
GEP - Glioepithelium
G/EP - Glioepithelium/ependyma
NEP - Neuroepithelium

3. Stem cells of choroid plexus
2. External germinal layer
1. Cerebellar NEP

Cerebellar germinal trigone

Settling Purkinje cells

Hemisphere
Lateral vermis
Medial vermis

CEREBELLUM (VERMIS)

Cerebellar NEP

Dorsal rhombic lip (cerebellar germinal trigone)

Rhombencephalic choroid plexus

CEREBELLUM (HEMISPHERE)

LATERAL POOL

RHOMBENCEPHALIC SUPERVENTRICLE (FUTURE FOURTH VENTRICLE)

MEDIAL POOL

Pontine (midline) G/EP

Pontine (facial) G/EP

Sprouting fibers of hook bundle?

Pontine (trigeminal) G/EP

Pontine (vestibulocochlear) G/EP

Trigeminal nerve fibers (V)

Genu of facial nerve fibers (VII)

Central gray

Cochlear nucleus (dorsal)

Lateral lemniscus

Principal sensory nucleus (V)

Trigeminal motor nucleus (V)

Reticular formation

Decussation of reticular formation fibers

Raphe nuclear complex

Cochlear nucleus (ventral)

PONS

External germinal layer (cochlear nuclei)

Ventral rhombic lip

Proximal nerve V (trigeminal)

Middle cerebellar peduncle

Medial lemniscus and trapezoid body

Pontocerebellar fibers

Pontine gray

Pontocerebellar fibers (decussation)

Reticular tegmental nucleus

Anterior extramural migratory stream
(pontine gray neurons intermingled with pioneer fibers of the middle cerebellar peduncle)

Note that corticofugal fibers are *absent* in the pontine gray.

Trigeminal ganglion (V)

Arrows indicate the presumed *direction of neuron migration* from neuroepithelial sources.

Arrows indicate the presumed *direction of axon growth* in nerves and fiber tracts.

Dashed lines indicate staining and/or sectioning artifacts.

42

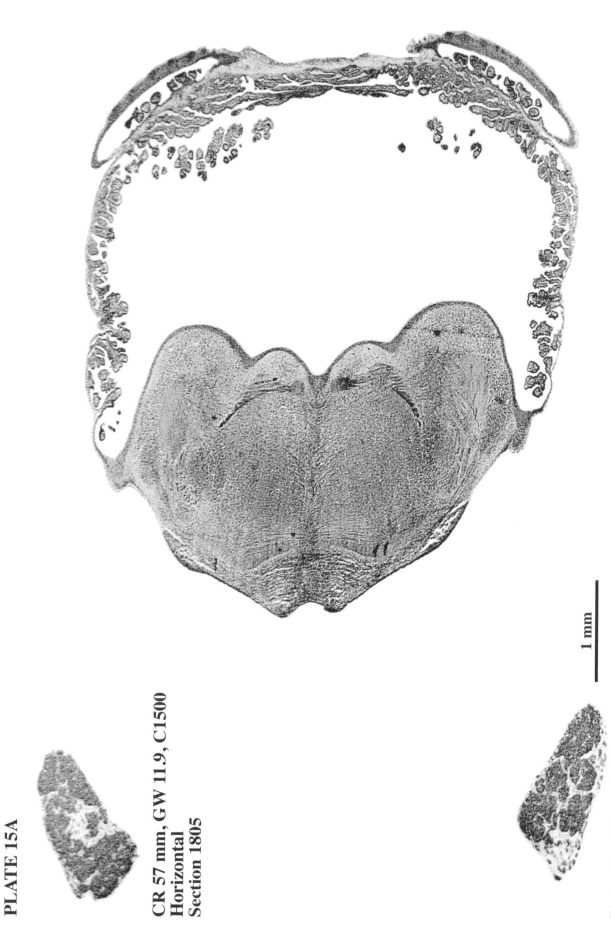

PLATE 15A

CR 57 mm, GW 11.9, C1500
Horizontal
Section 1805

1 mm

Plates 14 to 21 are shown at higher magnification.

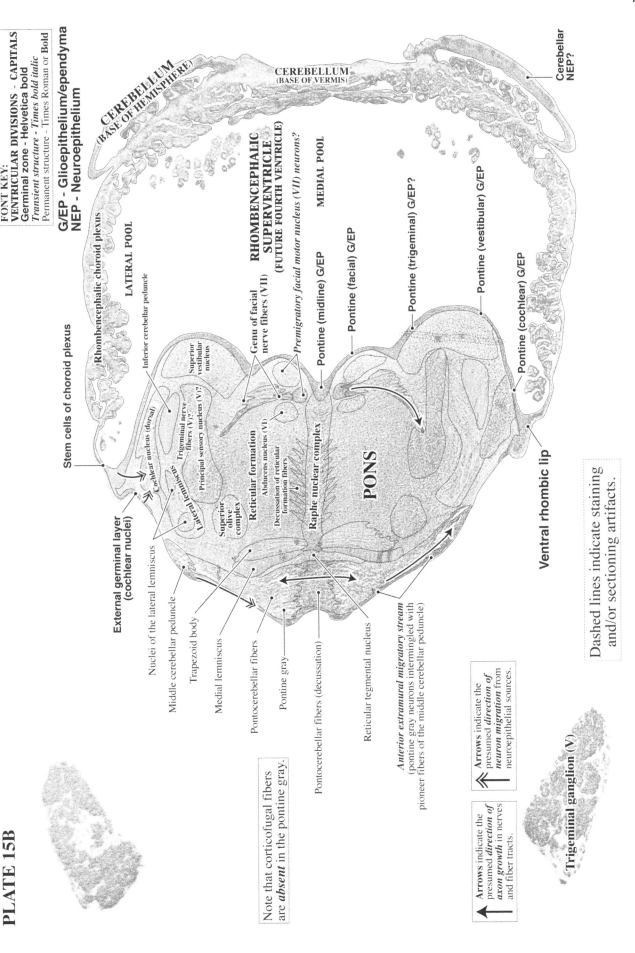

PLATE 15B

43

FONT KEY:
VENTRICULAR DIVISIONS - CAPITALS
Germinal zone - Helvetica bold
Transient structure - Times bold italic
Permanent structure - Times Roman or **Bold**
G/EP - Glioepithelium/ependyma
NEP - Neuroepithelium

CEREBELLUM
(BASE OF HEMISPHERE)

CEREBELLUM
(BASE OF VERMIS)

Cerebellar
NEP?

Rhombencephalic choroid plexus

LATERAL POOL

Inferior cerebellar peduncle

RHOMBENCEPHALIC
SUPERVENTRICLE
(FUTURE FOURTH VENTRICLE)

Premigratory facial motor nucleus (VII) neurons?

*Genu of facial
nerve fibers (VII)*

MEDIAL POOL

Superior
vestibular
nucleus

Pontine (midline) G/EP

Pontine (facial) G/EP

Pontine (trigeminal) G/EP?

Pontine (vestibular) G/EP

Pontine (cochlear) G/EP

Stem cells of choroid plexus

Cochlear nucleus (dorsal)

*Trigeminal nerve
fibers (V)?*

Principal sensory nucleus (V)?

Reticular formation

Abducens nucleus (VI)

*Decussation of reticular
formation fibers*

Raphe nuclear complex

PONS

External germinal layer
(cochlear nuclei)

Lateral lemniscus

Superior
olive
complex

Ventral rhombic lip

Nuclei of the lateral lemniscus

Middle cerebellar peduncle

Trapezoid body

Medial lemniscus

Pontocerebellar fibers

Pontine gray

Pontocerebellar fibers (decussation)

Reticular tegmental nucleus

Anterior extramural migratory stream
(pontine gray neurons intermingled with
pioneer fibers of the middle cerebellar peduncle)

Note that corticofugal fibers
are *absent* in the pontine gray.

Arrows indicate the
presumed *direction of
axon growth* in nerves
and fiber tracts.

Arrows indicate the
presumed *direction of
neuron migration* from
neuroepithelial sources.

Dashed lines indicate staining
and/or sectioning artifacts.

Trigeminal ganglion (V)

44

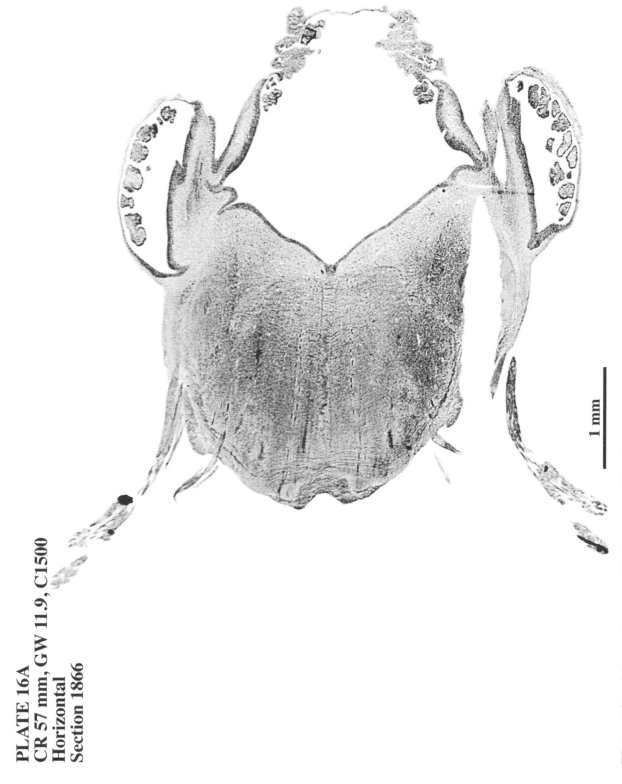

PLATE 16A
CR 57 mm, GW 11.9, C1500
Horizontal
Section 1866

1 mm

Plates 14 to 21 are shown at higher magnification.

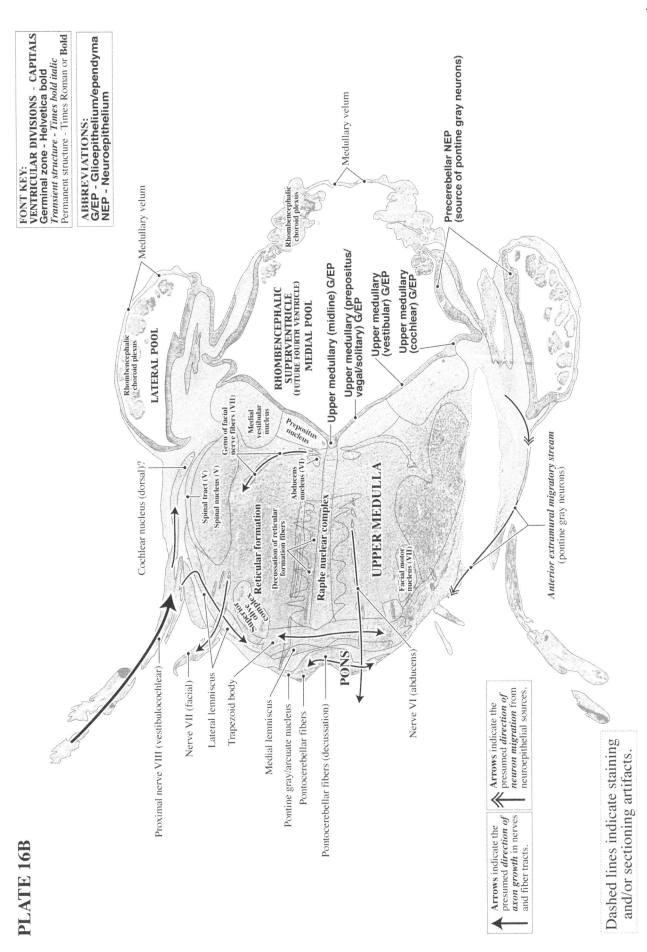

PLATE 16B

45

FONT KEY:
VENTRICULAR DIVISIONS - CAPITALS
Germinal zone - **Helvetica bold**
Transient structure - Times bold italic
Permanent structure - Times Roman or **Bold**

ABBREVIATIONS:
G/EP - Glioepithelium/ependyma
NEP - Neuroepithelium

Medullary velum

Precerebellar NEP
(source of pontine gray neurons)

Medullary velum

Rhombencephalic
choroid plexus

**RHOMBENCEPHALIC
SUPERVENTRICLE**
(FUTURE FOURTH VENTRICLE)
MEDIAL POOL

Rhombencephalic
choroid plexus

LATERAL POOL

Upper medullary (midline) G/EP

**Upper medullary (prepositus/
vagal/solitary) G/EP**

**Upper medullary
(vestibular) G/EP**

Upper medullary (cochlear) G/EP

Cochlear nucleus (dorsal)?

*Genu of facial
nerve fibers (VII)*

*Medial
vestibular
nucleus*

*Prepositus
nucleus*

Spinal tract (V)
Spinal nucleus (V)

*Abducens
nucleus (VI)*

Reticular formation

*Decussation of reticular
formation fibers*

Raphe nuclear complex

UPPER MEDULLA

*Superior
olive
complex*

*Facial motor
nucleus (VII)*

Anterior extramural migratory stream
(pontine gray neurons)

Proximal nerve VIII (vestibulocochlear)

Nerve VII (facial)

Lateral lemniscus

Trapezoid body

PONS

Nerve VI (abducens)

Medial lemniscus

Pontine gray/arcuate nucleus

Pontocerebellar fibers

Pontocerebellar fibers (decussation)

Arrows indicate the
presumed *direction of
axon growth* in nerves
and fiber tracts.

⇐ Arrows indicate the
presumed *direction of
neuron migration* from
neuroepithelial sources.

Dashed lines indicate staining
and/or sectioning artifacts.

46

PLATE 17A
CR 57 mm, GW 11.9, C1500
Horizontal
Section 1926

1 mm

Plates 14 to 21 are shown at higher magnification.

PLATE 17B

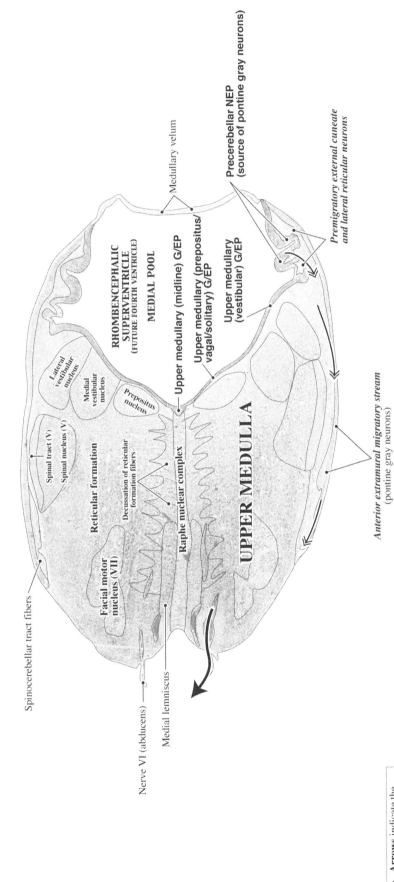

Medullary velum

Precerebellar NEP
(source of pontine gray neurons)

*Premigratory external cuneate
and lateral reticular neurons*

**RHOMBENCEPHALIC
SUPERVENTRICLE**
(FUTURE FOURTH VENTRICLE)

MEDIAL POOL

Upper medullary (midline) G/EP

Upper medullary (prepositus/
vagal/solitary) G/EP

Upper medullary
(vestibular) G/EP

*Lateral
vestibular
nucleus*

*Medial
vestibular
nucleus*

*Prepositus
nucleus*

Spinal tract (V)
Spinal nucleus (V)

Reticular formation

*Decussation of reticular
formation fibers*

Raphe nuclear complex

UPPER MEDULLA

*Facial motor
nucleus (VII)*

Anterior extramural migratory stream
(pontine gray neurons)

Spinocerebellar tract fibers

Nerve VI (abducens)

Medial lemniscus

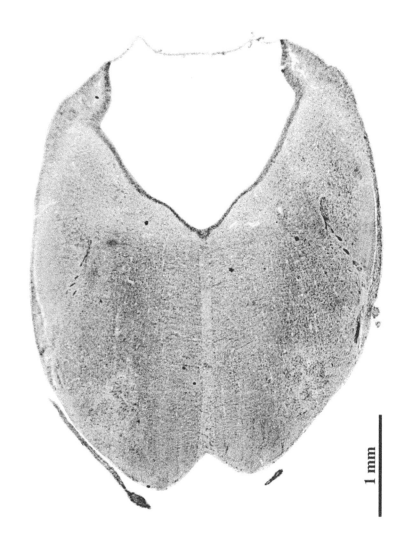

PLATE 18A
CR 57 mm, GW 11.9, C1500
Horizontal
Section 1962

1 mm

Plates 14 to 21 are shown at higher magnification.

PLATE 18B

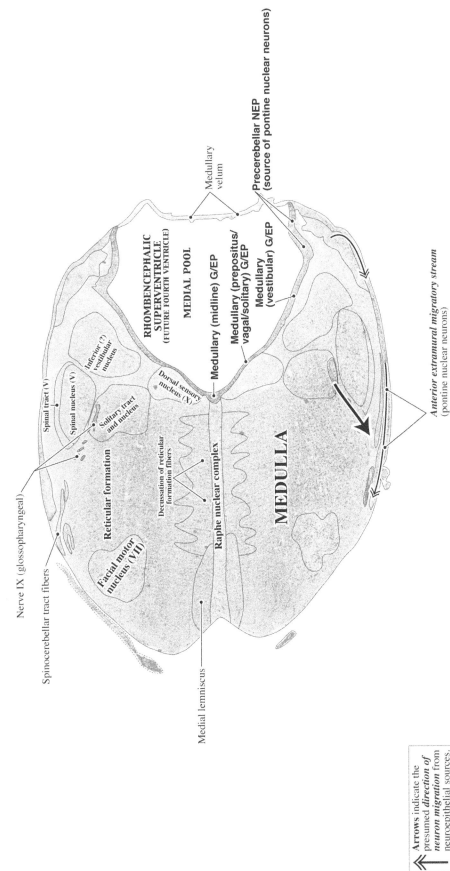

Medullary velum

Precerebellar NEP
(source of pontine nuclear neurons)

RHOMBENCEPHALIC
SUPERVENTRICLE
(FUTURE FOURTH VENTRICLE)

MEDIAL POOL

Medullary (midline) G/EP

Medullary (prepositus/
vagal/solitary) G/EP

Medullary
(vestibular) G/EP

Anterior extramural migratory stream
(pontine nuclear neurons)

Inferior (?)
vestibular
nucleus

Spinal tract (V)

Spinal nucleus (V)

Solitary tract
and nucleus

Dorsal sensory
nucleus (X)

Nerve IX (glossopharyngeal)

Spinocerebellar tract fibers

Reticular formation

Decussation of reticular
formation fibers

Raphe nuclear complex

Facial motor
nucleus (VII)

MEDULLA

Medial lemniscus

Dashed lines indicate staining
and/or sectioning artifacts.

Arrows indicate the
presumed *direction of*
neuron migration from
neuroepithelial sources.

Arrows indicate the
presumed *direction of*
axon growth in nerves
and fiber tracts.

PLATE 19A
CR 57 mm, GW 11.9, C1500
Horizontal
Section 2202

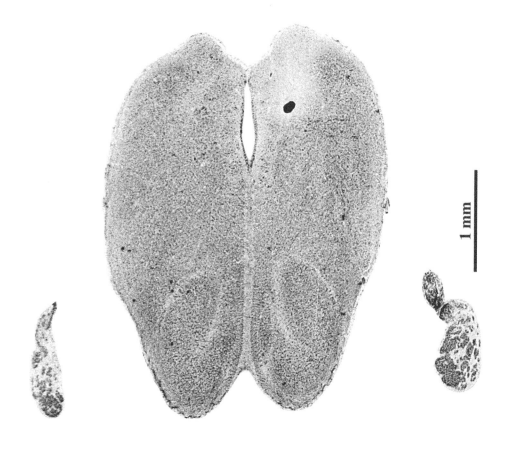

Plates 14 to 21 are shown at higher magnification.

1 mm

PLATE 19B

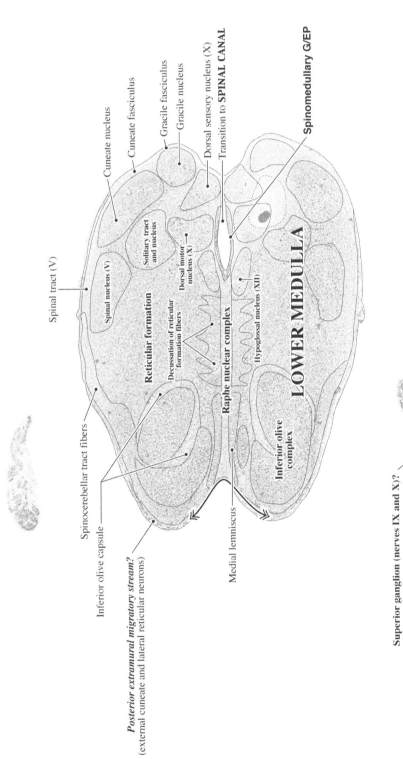

Cuneate nucleus

Cuneate fasciculus

Gracile fasciculus

Gracile nucleus

Dorsal sensory nucleus (X)

Transition to **SPINAL CANAL**

Spinomedullary G/EP

Spinal tract (V)

Spinal nucleus (V)

Solitary tract and nucleus

Reticular formation

Decussation of reticular formation fibers

Dorsal motor nucleus (X)

Raphe nuclear complex

Hypoglossal nucleus (XII)

LOWER MEDULLA

Inferior olive complex

Spinocerebellar tract fibers

Inferior olive capsule

Posterior extramural migratory stream?
(external cuneate and lateral reticular neurons)

Medial lemniscus

Superior ganglion (nerves IX and X)?

Dashed lines indicate staining and/or sectioning artifacts.

Arrows indicate the presumed *direction of neuron migration* from neuroepithelial sources.

52

PLATE 20A
CR 57 mm, GW 11.9, C1500
Horizontal
Section 2322

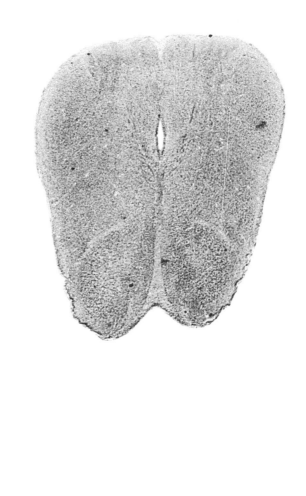

1 mm

Plates 14 to 21 are shown at higher magnification.

PLATE 20B

FONT KEY:
VENTRICULAR DIVISIONS - CAPITALS
Germinal zone - Helvetica bold
Transient structure - Times bold italic
Permanent structure - Times Roman or **Bold**

G/EP - Glioepithelium/ependyma

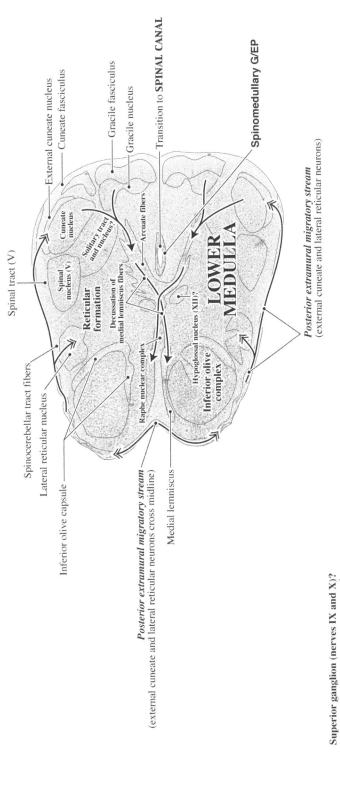

External cuneate nucleus

Cuneate fasciculus

Gracile fasciculus

Gracile nucleus

Transition to **SPINAL CANAL**

Spinomedullary G/EP

Posterior extramural migratory stream
(external cuneate and lateral reticular neurons)

Spinal tract (V)

Cuneate nucleus

Solitary tract and nucleus?

Arcuate fibers

Spinal nucleus (V)

Reticular formation

Decussation of medial lemniscus fibers

LOWER MEDULLA

Hypoglossal nucleus (XII)?

Inferior olive complex

Spinocerebellar tract fibers

Lateral reticular nucleus

Raphe nuclear complex

Inferior olive capsule

Posterior extramural migratory stream
(external cuneate and lateral reticular neurons cross midline)

Medial lemniscus

Superior ganglion (nerves IX and X)?

Inferior ganglion (nerves IX and X)?

⇐ **Arrows** indicate the presumed *direction of neuron migration* from neuroepithelial sources.

← **Arrows** indicate the presumed *direction of axon growth* in brain fiber tracts.

54

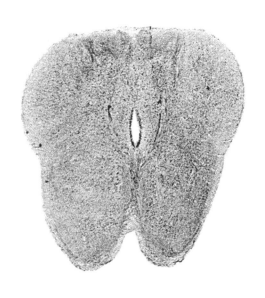

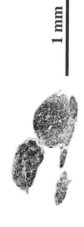

1 mm

PLATE 21A
CR 57 mm, GW 11.9, C1500
Horizontal
Section 2402

Plates 14 to 21 are shown at higher magnification.

PLATE 21B

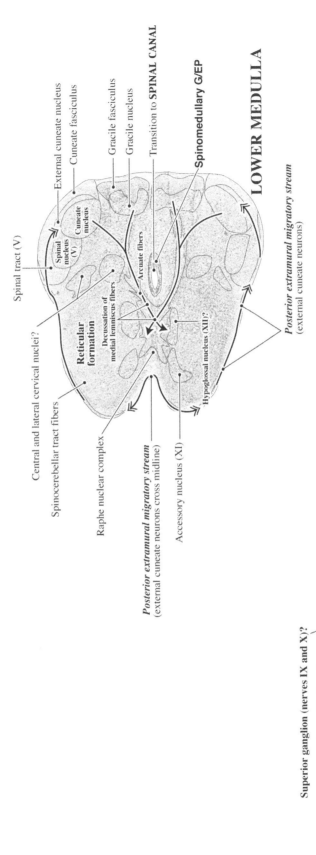

External cuneate nucleus

Cuneate fasciculus

Gracile fasciculus

Gracile nucleus

Transition to **SPINAL CANAL**

Spinomedullary G/EP

LOWER MEDULLA

Spinal tract (V)

Spinal nucleus (V)

Cuneate nucleus

Arcuate fibers

Reticular formation

Decussation of medial lemniscus fibers

Hypoglossal nucleus (XII)?

Posterior extramural migratory stream
(external cuneate neurons)

Central and lateral cervical nuclei?

Spinocerebellar tract fibers

Raphe nuclear complex

Posterior extramural migratory stream
(external cuneate neurons cross midline)

Accessory nucleus (XI)

Superior ganglion (nerves IX and X)?

Inferior ganglion (nerves IX and X)?

Arrows indicate the presumed *direction of **neuron migration*** from neuroepithelial sources.

Arrows indicate the presumed *direction of **axon growth*** in brain fiber tracts.

PART III: Y1-59
CR 60 mm (GW 12.5)
Frontal

This specimen is a stillborn female fetus with a crown rump length (CR) of 60 mm estimated to be at gestational week (GW) 12.5 (Yakovlev case number RPSL-WX-1-59, referred to here as Y1-59). The brain was cut in the coronal (frontal) plane in 35-μm-thick sections and is classified as a Normative Control in the Yakovlev Collection (Haleem, 1990). Since there is no photograph of this brain before it was embedded and cut, a specimen from Hochstetter (1919) that is comparable to Y1-59 is used to show the approximate section plane and external features of a GW 12.5 brain (**Figure 9**). Photographs of 22 Nissl-stained sections are shown at low magnification in **Plates 22–42**. **Plates 43–56** show high-magnification views of various parts of the brain from the cerebral cortex to the midbrain, pons, and medulla. Several high-magnification plates are rotated 90° (landscape orientation) to more efficiently use page space. Y1-59 has more mature brain structures in the diencephalon, midbrain, pons, and medulla than any other specimen in the first trimester. Immature brain structures predominate in the telencephalon and the cerebellum.

Throughout the cerebral cortex, the neuroepithelium and subventricular zone are prominent. The stratified transitional field (STF) contains mainly STF1 and STF5 throughout, STF4 only in lateral areas and a questionable STF2 in a few areas; STF6 and STF3 are not present. The STF is filled with migrating and sojourning neurons and has no regional heterogeneity. The cerebral cortex is completely smooth except for a questionable calcarine sulcus in the left cerebral hemisphere. The most prominent developmental feature of the cerebral cortex is that both the STF layers and the cortical plate have a pronounced lateral (thicker) to medial (thinner) maturation gradient. In anterolateral parts of the cerebral cortex, streams of neurons and glia appear to leave STF4 and enter the lateral migratory stream. There is no corpus callosum, but fibers are approaching the midline. The olfactory bulb is beneath the anterior septum; it contains the rostral migratory stream in its core. The hippocampus is in an immature position dorsal to the thalamus. Cells are entering Ammon's horn pyramidal layer in the ammonic migration, and granule cells and their precursors are migrating to the hilus of the presumptive dentate gyrus in the dentate migration; there is no granular layer. A massive neuroepithelium/subventricular zone overlies the nucleus accumbens and striatum (caudate and putamen) where neurons (and glia) are being generated. The strionuclear glioepithelium forms definite continuities with the fornical glioepithelium in the telencephalon.

The cerebellum is a thick, smooth plate overlying the posterior pons and medulla. However, there is only a thin neuroepithelium at the ventricular surface, possibly generating Golgi cells that will disperse in the cerebellar granule cell layer. The deep neurons are in place beneath the primordial cortex, but have indefinite nuclear subdivisions. The cortical surface is covered by an external germinal layer (egl) that is actively producing neuronal stem cells (granule, stellate, and basket cells) of the cerebellar cortex. Lamination in the cortex is nearly absent, except for a thin molecular layer beneath the egl. Nearly all Purkinje cells are migrating, some in discrete clumps. Lobulation has barely begun in the vermis and is nearly absent in the hemispheres. The germinal trigone is prominent in the dorsal rhombic lip.

The third ventricle, aqueduct, and fourth ventricle are lined by a thin glioepithelium/ependyma indicating that neurogenesis in the primary neuroepithelium is complete. There are a few exceptions: the arcuate neuroepithelium is still active, the inferior collicular neuroepithelium is generating neurons and the precerebellar neuroepithelium in the medulla is generating pontine gray neurons.

Neurons throughout the diencephalon are settling in fairly well-defined nuclear divisions; the major exceptions are the immature appearance of the lateral and medial geniculate bodies in the posterior thalamus and the hypothalamic medial mammillary body. Neurons are settled in the midbrain tegmentum, pons, and medulla. But the pontine gray is small; neurons are still migrating into it from the large anterior extramural migratory stream. Remnants of the posterior extramural and intramural migratory streams are tentatively identified. The corticospinal tract forms a small cerebral peduncle in the midbrain, and has just penetrated the pons.

GW 12.5 CORONAL SECTION PLANES

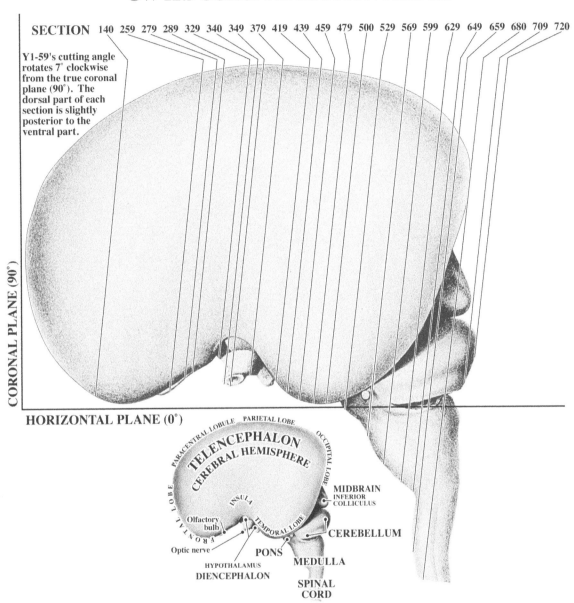

SECTION 140 259 279 289 329 340 349 379 419 439 459 479 500 529 569 599 629 649 659 680 709 720

Y1-59's cutting angle rotates 7° clockwise from the true coronal plane (90°). The dorsal part of each section is slightly posterior to the ventral part.

CORONAL PLANE (90°)

HORIZONTAL PLANE (0°)

PARACENTRAL LOBULE   PARIETAL LOBE
TELENCEPHALON
CEREBRAL HEMISPHERE
OCCIPITAL LOBE
FRONTAL LOBE
INSULA
MIDBRAIN
INFERIOR COLLICULUS
Olfactory bulb
TEMPORAL LOBE
CEREBELLUM
Optic nerve
PONS
HYPOTHALAMUS
MEDULLA
DIENCEPHALON
SPINAL CORD

Figure 9. The lateral view of the brain and upper cervical spinal cord from a specimen with a crown rump length of 68 mm (modified from Figure 47, Table VIII, Hochstetter, 1919) serves to show the approximate locations and cutting angles of the illustrated sections of Y1-59 in the following pages. The small inset identifies the major structural features. The cut beneath the cerebellum is the edge of the medullary velum.

PLATE 22A
CR 60 mm, GW 12.5, Y1-59
Frontal
Section 140

LAYERS OF THE CORTICAL
STRATIFIED TRANSITIONAL FIELD (STF)

STF1   Superficial fibrous layer with an early developmental stage *(t1)* when many cells are migrating through it, followed by a late stage *(t2)* with sparse cells. Endures as the subcortical white matter.

STF4   Complex middle layer where sojourning and migrating cortical neurons grow corticofugal axons and intermingle with corticopetal axons.

STF5   Deep cellular layer that is prominent during the first trimester, the first sojourn zone to appear outside the germinal matrix.

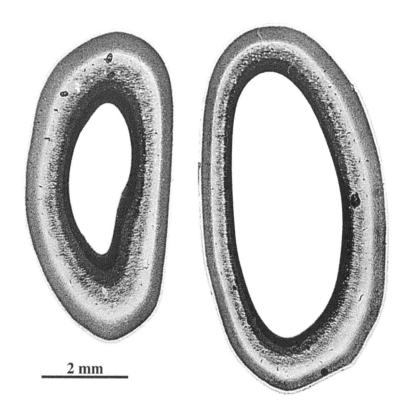

2 mm

FONT KEY:
VENTRICULAR DIVISIONS - CAPITALS
Germinal zone - Helvetica bold
Transient structure - Times bold italic
Permanent structure - Times Roman or **Bold**

ABBREVIATIONS:
NEP - Neuroepithelium
SVZ - Subventricular zone

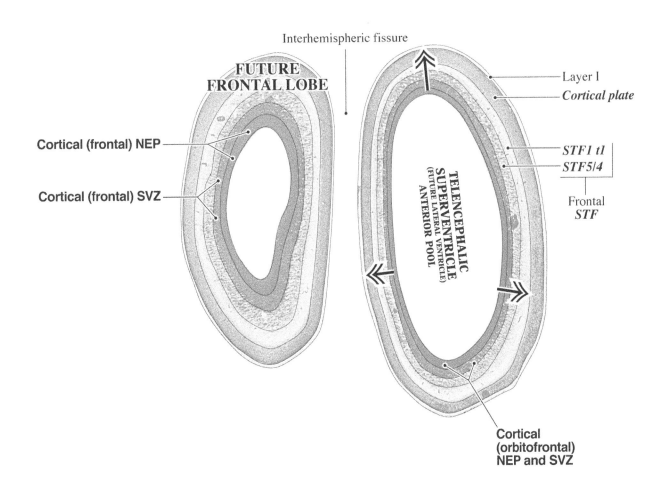

Interhemispheric fissure

FUTURE
FRONTAL LOBE

Layer I

Cortical plate

Cortical (frontal) NEP

STF1 t1
STF5/4

Cortical (frontal) SVZ

Frontal
STF

TELENCEPHALIC
SUPERVENTRICLE
(FUTURE LATERAL VENTRICLE)
ANTERIOR POOL

Cortical
(orbitofrontal)
NEP and SVZ

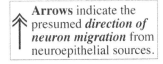

Arrows indicate the
presumed *direction of
neuron migration* from
neuroepithelial sources.

PLATE 23A
CR 60 mm, GW 12.5, Y1-59
Frontal
Section 259

LAYERS OF THE CORTICAL
STRATIFIED TRANSITIONAL
FIELD (STF)

STF1 Superficial fibrous layer with an early developmental stage *(t1)* when many cells are migrating through it, followed by a late stage *(t2)* with sparse cells. Endures as the subcortical white matter.

STF2 Upper cellular layer, the most superficial sojourn zone where cells translocate to the cortical plate.

STF4 Complex middle layer where sojourning and migrating cortical neurons grow corticofugal axons and intermingle with corticopetal axons.

STF5 Deep cellular layer that is prominent during the first trimester, the first sojourn zone to appear outside the germinal matrix.

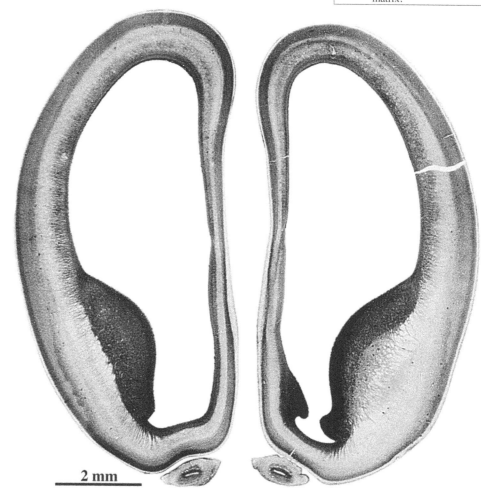

2 mm

See a high-magnification view of the frontal cortex
from Section 269 in Plates 44A and B.

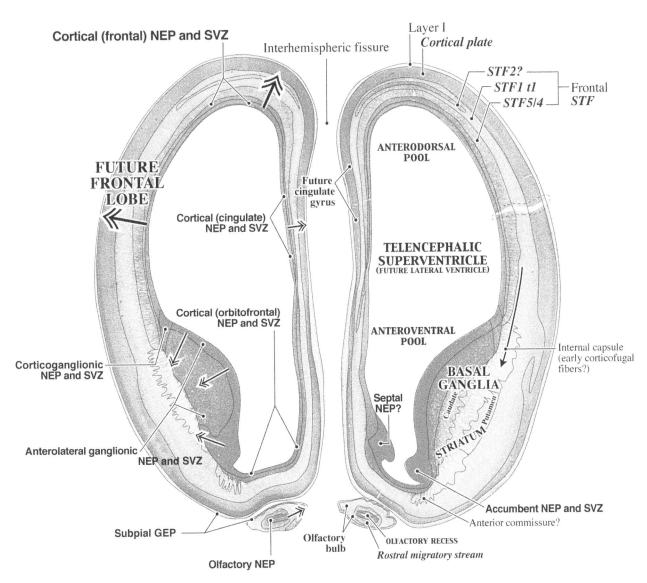

Cortical (frontal) NEP and SVZ

Interhemispheric fissure

Layer I
Cortical plate

STF2?
STF1 t1
STF5/4

Frontal
STF

**ANTERODORSAL
POOL**

**FUTURE
FRONTAL
LOBE**

*Future
cingulate
gyrus*

**Cortical (cingulate)
NEP and SVZ**

**TELENCEPHALIC
SUPERVENTRICLE**
(FUTURE LATERAL VENTRICLE)

**Cortical (orbitofrontal)
NEP and SVZ**

**ANTEROVENTRAL
POOL**

Internal capsule
(early corticofugal
fibers?)

**Corticoganglionic
NEP and SVZ**

**BASAL
GANGLIA**

*Septal
NEP?*

Caudate

Putamen

STRIATUM

**Anterolateral ganglionic
NEP and SVZ**

Subpial GEP

Accumbent NEP and SVZ
Anterior commissure?

**Olfactory
bulb**

OLFACTORY RECESS
Rostral migratory stream

Olfactory NEP

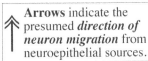

Arrows indicate the
presumed *direction of
neuron migration* from
neuroepithelial sources.

Arrows indicate the
presumed *direction of
axon growth* in brain
fiber tracts.

Dashed lines indicate staining
and/or sectioning artifacts.

**PLATE 24A
CR 60 mm, GW 12.5, Y1-59
Frontal
Section 279**

LAYERS OF THE CORTICAL
*STRATIFIED TRANSITIONAL
FIELD (STF)*

STF1	Superficial fibrous layer with an early developmental stage *(t1)* when many cells are migrating through it, followed by a late stage *(t2)* with sparse cells. Endures as the subcortical white matter.
STF2	Upper cellular layer, the most superficial sojourn zone where cells translocate to the cortical plate.
STF4	Complex middle layer where sojourning and migrating cortical neurons grow corticofugal axons and intermingle with corticopetal axons.
STF5	Deep cellular layer that is prominent during the first trimester, the first sojourn zone to appear outside the germinal matrix.

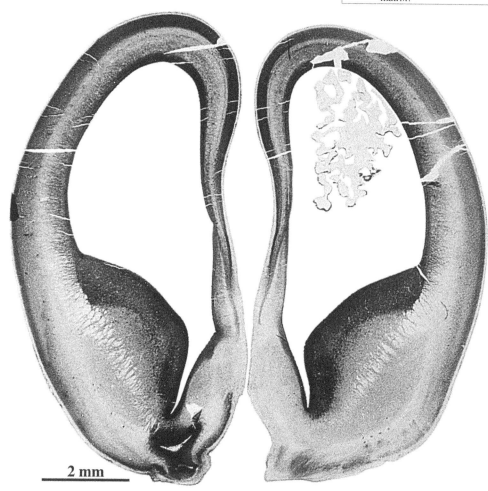

2 mm

**See a high-magnification view of the frontal cortex
from Section 269 in Plates 44A and B.**

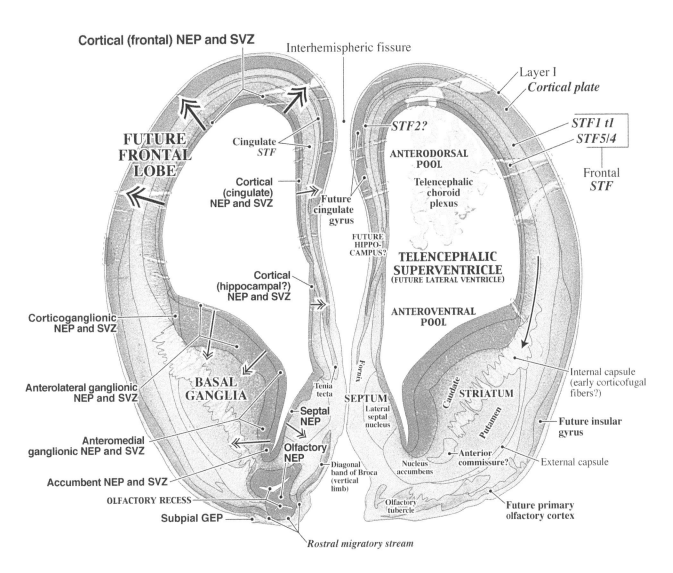

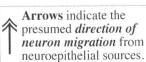

Arrows indicate the presumed *direction of neuron migration* from neuroepithelial sources.

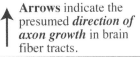

Arrows indicate the presumed *direction of axon growth* in brain fiber tracts.

Dashed lines indicate staining and/or sectioning artifacts.

PLATE 25A
CR 60 mm, GW 12.5, Y1-59
Frontal
Section 289

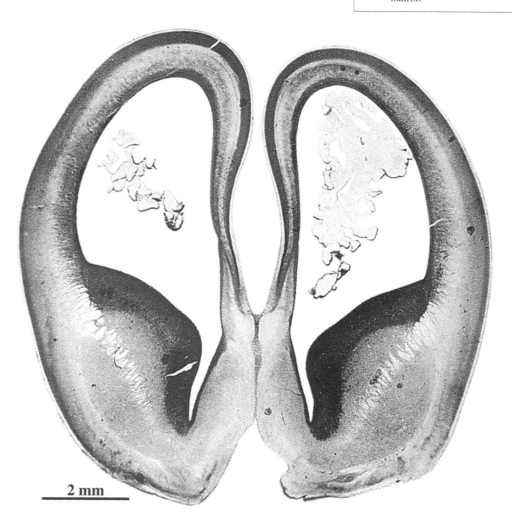

2 mm

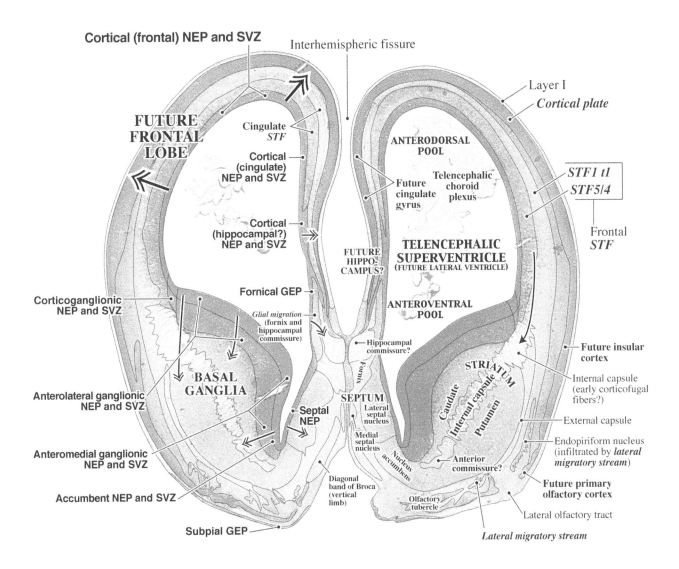

FONT KEY:
VENTRICULAR DIVISIONS - CAPITALS
Germinal zone - Helvetica bold
Transient structure - Times bold italic
Permanent structure - Times Roman or **Bold**

ABBREVIATIONS:
GEP - Glioepithelium
NEP - Neuroepithelium
SVZ - Subventricular zone

Cortical (frontal) NEP and SVZ

Interhemispheric fissure

FUTURE FRONTAL LOBE

Cingulate *STF*

Cortical (cingulate) NEP and SVZ

Cortical (hippocampal?) NEP and SVZ

Corticoganglionic NEP and SVZ

Anterolateral ganglionic NEP and SVZ

BASAL GANGLIA

Anteromedial ganglionic NEP and SVZ

Accumbent NEP and SVZ

Subpial GEP

Fornical GEP

Glial migration (fornix and hippocampal commissure)

FUTURE HIPPO-CAMPUS?

Septal NEP

SEPTUM

Lateral septal nucleus

Medial septal nucleus

Diagonal band of Broca (vertical limb)

Fornix

Hippocampal commissure?

Nucleus accumbens

Olfactory tubercle

Layer I
Cortical plate

ANTERODORSAL POOL

Future cingulate gyrus

Telencephalic choroid plexus

STF1 t1
STF5/4

Frontal *STF*

TELENCEPHALIC SUPERVENTRICLE (FUTURE LATERAL VENTRICLE)

ANTEROVENTRAL POOL

STRIATUM

Caudate

Internal capsule

Putamen

Anterior commissure?

Future insular cortex

Internal capsule (early corticofugal fibers?)

External capsule

Endopiriform nucleus (infiltrated by *lateral migratory stream*)

Future primary olfactory cortex

Lateral olfactory tract

Lateral migratory stream

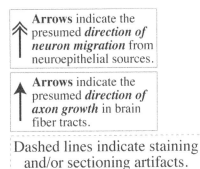

Arrows indicate the presumed *direction of neuron migration* from neuroepithelial sources.

Arrows indicate the presumed *direction of axon growth* in brain fiber tracts.

Dashed lines indicate staining and/or sectioning artifacts.

PLATE 26A
CR 60 mm, GW 12.5, Y1-59
Frontal
Section 329

LAYERS OF THE CORTICAL
STRATIFIED TRANSITIONAL FIELD (STF)

STF1　Superficial fibrous layer with an early developmental stage *(t1)* when many cells are migrating through it, followed by a late stage *(t2)* with sparse cells. Endures as the subcortical white matter.

STF4　Complex middle layer where sojourning and migrating cortical neurons grow corticofugal axons and intermingle with corticopetal axons.

STF5　Deep cellular layer that is prominent during the first trimester, the first sojourn zone to appear outside the germinal matrix.

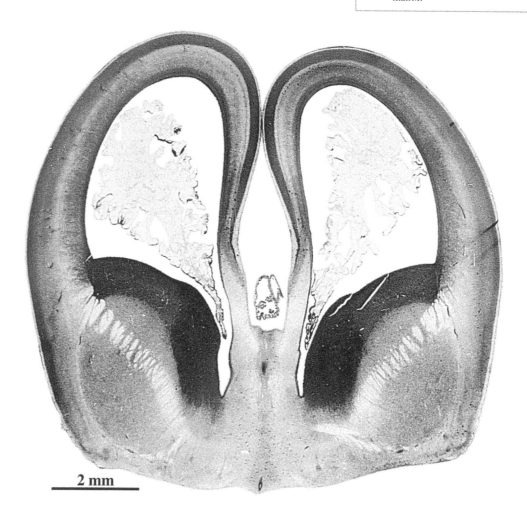

2 mm

FONT KEY:
VENTRICULAR DIVISIONS - CAPITALS
Germinal zone - Helvetica bold
Transient structure - Times bold italic
Permanent structure - Times Roman or **Bold**

ABBREVIATIONS:
GEP - Glioepithelium
G/EP - Glioepithelium/ependyma
NEP - Neuroepithelium
SVZ - Subventricular zone

Cortical (frontal) NEP and SVZ

Interhemispheric fissure

Layer I
Cortical plate

FUTURE FRONTAL LOBE

Cingulate *STF*

Cortical (cingulate) NEP and SVZ

Cortical (hippocampal) NEP and SVZ

Ammonic migration

Dentate migration

Fornical GEP

Corticoganglionic NEP and SVZ

Anterolateral ganglionic NEP and SVZ

Anteromedial ganglionic NEP and SVZ

Accumbent NEP and SVZ

Subpial GEP

Preoptic G/EP

ANTERODORSAL POOL

Future cingulate gyrus

Telencephalic choroid plexus

HIPPO-CAMPUS

TELENCEPHALIC SUPERVENTRICLE
(FUTURE LATERAL VENTRICLE)

ANTEROVENTRAL POOL

Diencephalic choroid plexus

Fornix

Hippocampal commissure?

Septal NEP

Lateral septal nucleus

SEPTUM

Medial septal nucleus

BASAL GANGLIA

Nucleus accumbens

Anterior commissure

STRIATUM

Caudate

Internal capsule

Putamen

Substantia innominata

Medial forebrain bundle

Diagonal band of Broca (vertical limb)

Preoptic area

STF1 t1
STF4?
STF5/4

Frontal *STF*

Lateral migratory stream?

Internal capsule (early corticofugal fibers?)

External capsule
Future insular cortex

Claustrum

Lateral migratory stream

Endopiriform nucleus

Lateral olfactory tract

Future primary olfactory cortex

Arrows indicate the presumed *direction of neuron migration* from neuroepithelial sources.

Arrows indicate the presumed *direction of axon growth* in brain fiber tracts.

Dashed lines indicate staining and/or sectioning artifacts.

68

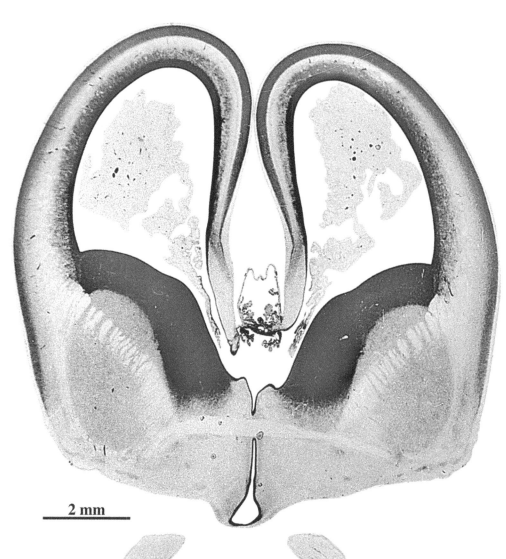

2 mm

LAYERS OF THE CORTICAL
STRATIFIED TRANSITIONAL FIELD (STF)

STF1 Superficial fibrous layer with an early
developmental stage *(t1)* when many
cells are migrating through it, followed
by a late stage *(t2)* with sparse cells.
Endures as the subcortical white matter.

STF4 Complex middle layer where sojourning
and migrating cortical neurons grow
corticofugal axons and intermingle with
corticopetal axons.

STF5 Deep cellular layer that is prominent
during the first trimester, the first sojourn
zone to appear outside the germinal
matrix.

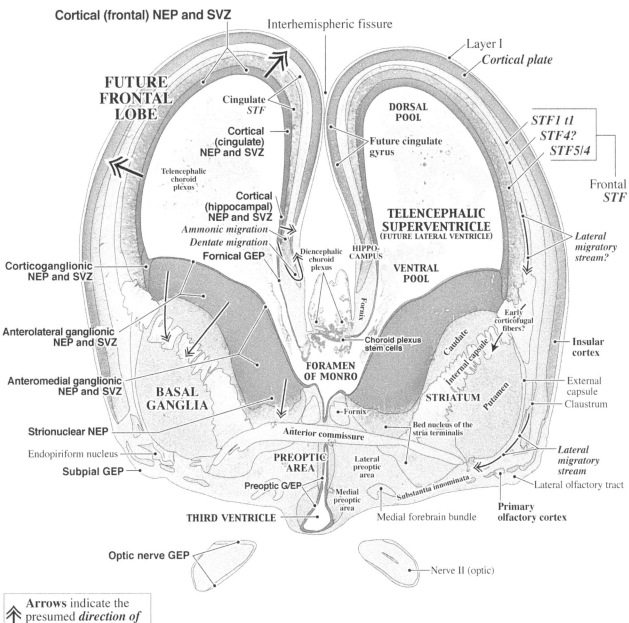

Cortical (frontal) NEP and SVZ

Interhemispheric fissure

Layer I
Cortical plate

**FUTURE
FRONTAL
LOBE**

Cingulate
STF

**DORSAL
POOL**

Cortical
(cingulate)
NEP and SVZ

Future cingulate
gyrus

*STF1 t1
STF4?
STF5/4*

Telencephalic
choroid
plexus

Frontal
STF

Cortical
(hippocampal)
NEP and SVZ
Ammonic migration
Dentate migration
Fornical GEP

**TELENCEPHALIC
SUPERVENTRICLE**
(FUTURE LATERAL VENTRICLE)

*Lateral
migratory
stream?*

Corticoganglionic
NEP and SVZ

Diencephalic
choroid
plexus

HIPPO-
CAMPUS

**VENTRAL
POOL**

Fornix

Early
corticofugal
fibers?

Anterolateral ganglionic
NEP and SVZ

Choroid plexus
stem cells

Caudate

internal capsule

**Insular
cortex**

Anteromedial ganglionic
NEP and SVZ

**FORAMEN
OF MONRO**

STRIATUM

Putamen

External
capsule

Claustrum

**BASAL
GANGLIA**

Fornix

Bed nucleus of the
stria terminalis

Strionuclear NEP

Anterior commissure

*Lateral
migratory
stream*

Endopiriform nucleus

**PREOPTIC
AREA**

Lateral
preoptic
area

Lateral olfactory tract

Subpial GEP

Preoptic G/EP

Substantia innominata

Medial
preoptic
area

**Primary
olfactory cortex**

THIRD VENTRICLE

Medial forebrain bundle

Optic nerve GEP

Nerve II (optic)

PLATE 28A
CR 60 mm, GW 12.5, Y1-59
Frontal
Section 349

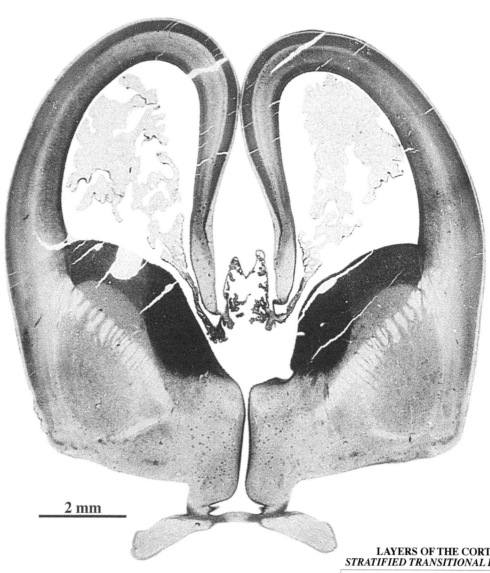

2 mm

LAYERS OF THE CORTICAL
STRATIFIED TRANSITIONAL FIELD (STF)

STF1   Superficial fibrous layer with an early
developmental stage *(t1)* when many
cells are migrating through it, followed
by a late stage *(t2)* with sparse cells.
Endures as the subcortical white matter.

STF4   Complex middle layer where sojourning
and migrating cortical neurons grow
corticofugal axons and intermingle with
corticopetal axons.

STF5   Deep cellular layer that is prominent
during the first trimester, the first sojourn
zone to appear outside the germinal
matrix.

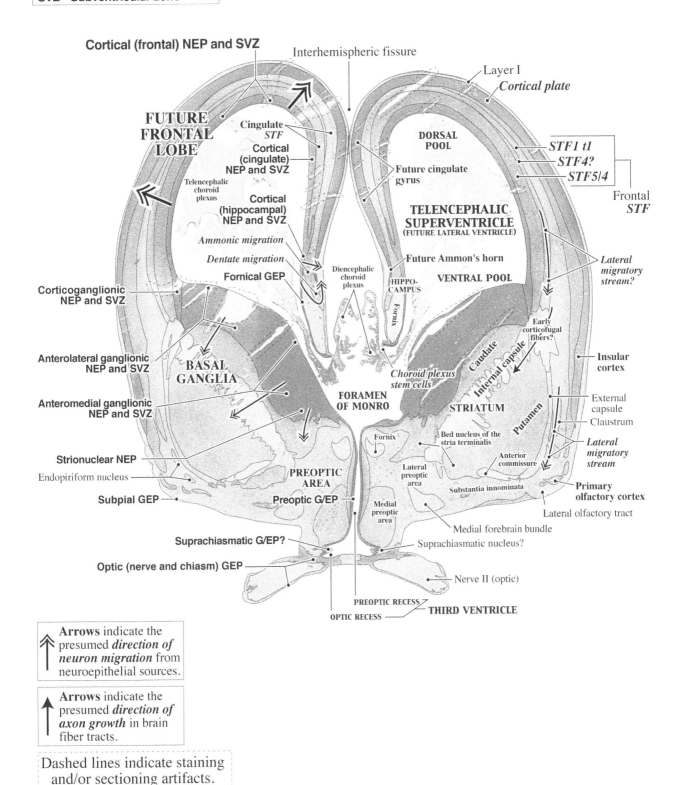

Arrows indicate the presumed *direction of neuron migration* from neuroepithelial sources.

Arrows indicate the presumed *direction of axon growth* in brain fiber tracts.

Dashed lines indicate staining and/or sectioning artifacts.

PLATE 29A
CR 60 mm, GW 12.5, Y1-59
Frontal
Section 379

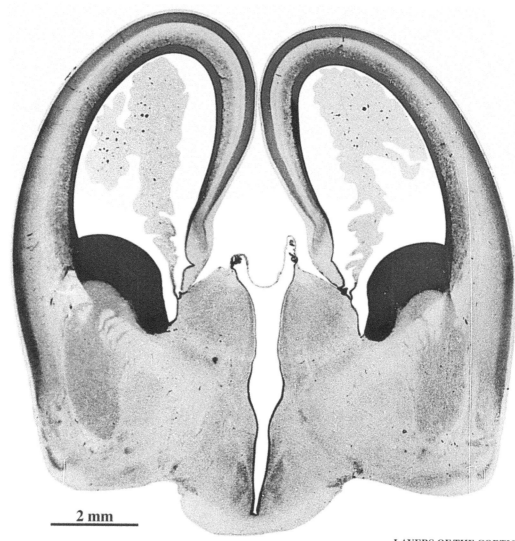

2 mm

See high-magnification views of the
right paracentral cortex, thalamus,
and basal ganglia from Section 399
in Plates 45 to 46A and B.

See a high-magnification view of
the diencephalon and basal
telencephalon from Section 389 in
Plates 47A and B.

LAYERS OF THE CORTICAL
STRATIFIED TRANSITIONAL FIELD (STF)

STF1 Superficial fibrous layer with an early
developmental stage *(t1)* when many
cells are migrating through it, followed
by a late stage *(t2)* with sparse cells.
Endures as the subcortical white matter.

STF4 Complex middle layer where sojourning
and migrating cortical neurons grow
corticofugal axons and intermingle with
corticopetal axons.

STF5 Deep cellular layer that is prominent
during the first trimester, the first sojourn
zone to appear outside the germinal
matrix.

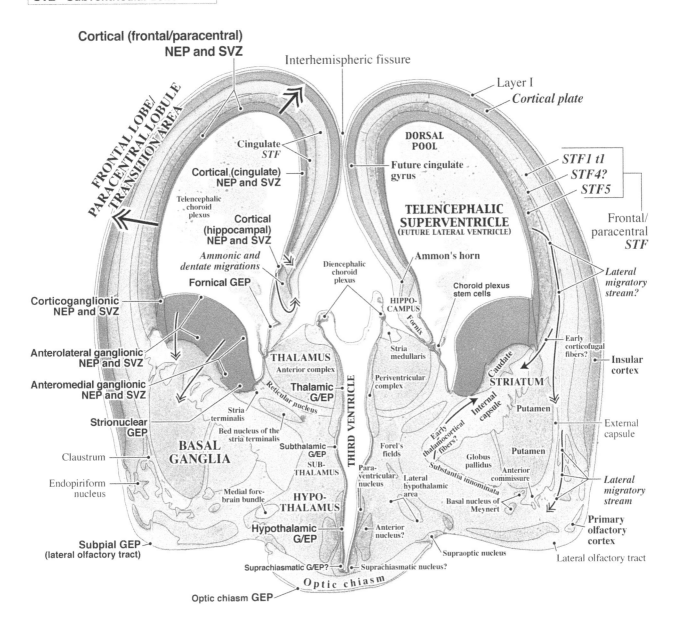

FONT KEY:
VENTRICULAR DIVISIONS - CAPITALS
Germinal zone - Helvetica bold
Transient structure - Times bold italic
Permanent structure - Times Roman or **Bold**

ABBREVIATIONS:
GEP - Glioepithelium
G/EP - Glioepithelium/ependyma
NEP - Neuroepithelium
SVZ - Subventricular zone

Cortical (frontal/paracentral)
NEP and SVZ

Interhemispheric fissure

Layer I
Cortical plate

FRONTAL LOBE/
PARACENTRAL LOBULE
TRANSITION AREA

Cingulate
STF

Cortical (cingulate)
NEP and SVZ

Telencephalic
choroid
plexus

Cortical
(hippocampal)
NEP and SVZ

*Ammonic and
dentate migrations*

Fornical GEP

Corticoganglionic
NEP and SVZ

Anterolateral ganglionic
NEP and SVZ

Anteromedial ganglionic
NEP and SVZ

Strionuclear
GEP

Claustrum

Endopiriform
nucleus

Subpial GEP
(lateral olfactory tract)

DORSAL
POOL

Future cingulate
gyrus

STF1 t1
STF4?
STF5

Frontal/
paracentral
STF

**TELENCEPHALIC
SUPERVENTRICLE**
(FUTURE LATERAL VENTRICLE)

*Lateral
migratory
stream?*

Ammon's horn

Diencephalic
choroid
plexus

Choroid plexus
stem cells

HIPPO-
CAMPUS

Fornix

THALAMUS
Anterior complex

Thalamic
G/EP

Reticular nucleus

Stria
terminalis

Bed nucleus of the
stria terminalis

BASAL
GANGLIA

Subthalamic
G/EP
SUB-
THALAMUS

Medial fore-
brain bundle

HYPO-
THALAMUS

Hypothalamic
G/EP

Suprachiasmatic G/EP?

Optic chiasm GEP

Stria
medullaris

Periventricular
complex

Forel's
fields

Para-
ventricular
nucleus

THIRD VENTRICLE

Early
thalamocortical
fibers?

Lateral
hypothalamic
area

Substantia innominata

Globus
pallidus

Basal nucleus of
Meynert

Anterior
nucleus?

Suprachiasmatic nucleus?

Optic chiasm

Supraoptic nucleus

*Early
corticofugal
fibers?*

Caudate

STRIATUM

Internal
capsule

Putamen

Putamen

Anterior
commissure

Insular
cortex

External
capsule

*Lateral
migratory
stream*

**Primary
olfactory
cortex**

Lateral olfactory tract

⇑ **Arrows** indicate the
presumed *direction of
neuron migration* from
neuroepithelial sources.

↑ **Arrows** indicate the
presumed *direction of
axon growth* in brain
fiber tracts.

PLATE 30A
CR 60 mm, GW 12.5, Y1-59
Frontal
Section 419

LAYERS OF THE CORTICAL
STRATIFIED TRANSITIONAL
FIELD (STF)

STF1 Superficial fibrous layer with an early developmental stage *(t1)* when many cells are migrating through it, followed by a late stage *(t2)* with sparse cells. Endures as the subcortical white matter.

STF2 Upper cellular layer, the most superficial sojourn zone where cells translocate to the cortical plate.

STF4 Complex middle layer where sojourning and migrating cortical neurons grow corticofugal axons and intermingle with corticopetal axons.

STF5 Deep cellular layer that is prominent during the first trimester, the first sojourn zone to appear outside the germinal matrix.

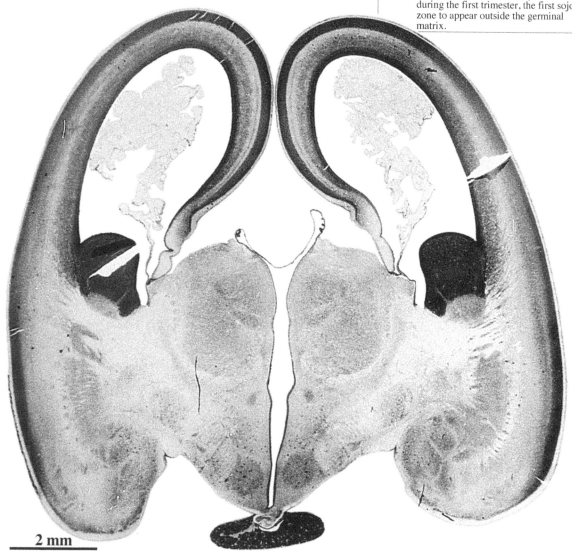

2 mm

See a high-magnification view of the diencephalon and basal telencephalon from this Section in Plates 48A and B.

Cortical (paracentral)
NEP and SVZ

Interhemispheric fissure

Layer I
Cortical plate

Cingulate
STF

DORSAL
POOL

FUTURE
PARACENTRAL
LOBULE

Cortical (cingulate)
NEP and SVZ

Future cingulate
gyrus

STF1 tl
STF4?
STF5/4

Telencephalic choroid
plexus

STF2?

Paracentral
STF

Cortical (hippocampal)
NEP and SVZ

TELENCEPHALIC
SUPERVENTRICLE
(FUTURE LATERAL VENTRICLE)

Ammonic and
dentate migrations

Future Ammon's horn

Fornical GEP

HIPPO-
CAMPUS

Diencephalic
choroid
plexus

Choroid plexus
stem cells

Lateral
migratory
stream?

Dorso-
lateral
nucleus

Corticoganglionic
NEP and SVZ

Dorso-
medial
nucleus

Stria
medullaris

Early thalamocortical
fibers?

THALAMUS

Dorsal complex

Posterior ganglionic
NEP and SVZ

Central
complex

Periventricular
complex

Caudate

STRIATUM

Insular
cortex

Strionuclear
GEP

Reticular nucleus

Ventral
complex

Internal
capsule

Stria
terminalis

External
capsule

FUTURE
TEMPORAL
LOBE

BASAL
GANGLIA

Zona incerta

Putamen

SUBTHALAMUS

THIRD VENTRICLE

Forel's fields

Globus pallidus

Subthalamic
G/EP

Ventral
striatum

Subpial
GEP

Medial fore-
brain bundle

Dorsomedial
nucleus

Cerebral
peduncle

Central
nucleus

HYPO-
THALAMUS

Corticomedial complex

Basolateral
complex

AMYGDALA

Optic tract

Hypothalamic
G/EP

Lateral tuberal
nucleus?

Endipiriform
nucleus

Ventromedial
nucleus

Arcuate nucleus

Lateral
migratory
stream?

Perifascicular
GEP

Median eminence/
posterior pituitary gland
(neurohypophysis)

Primary
olfactory cortex

Lateral olfactory tract

Anterior pituitary gland
(adenohypophysis, distal part)

Arrows indicate the
presumed *direction of*
neuron migration from
neuroepithelial sources.

Arrows indicate the
presumed *direction of*
axon growth in brain
fiber tracts.

Dashed lines indicate staining
and/or sectioning artifacts.

PLATE 31A
CR 60 mm, GW 12.5, Y1-59
Frontal
Section 439

LAYERS OF THE CORTICAL
STRATIFIED TRANSITIONAL
FIELD (STF)

STF1   Superficial fibrous layer with an early developmental stage *(t1)* when many cells are migrating through it, followed by a late stage *(t2)* with sparse cells. Endures as the subcortical white matter.

STF2   Upper cellular layer, the most superficial sojourn zone where cells translocate to the cortical plate.

STF4   Complex middle layer where sojourning and migrating cortical neurons grow corticofugal axons and intermingle with corticopetal axons.

STF5   Deep cellular layer that is prominent during the first trimester, the first sojourn zone to appear outside the germinal matrix.

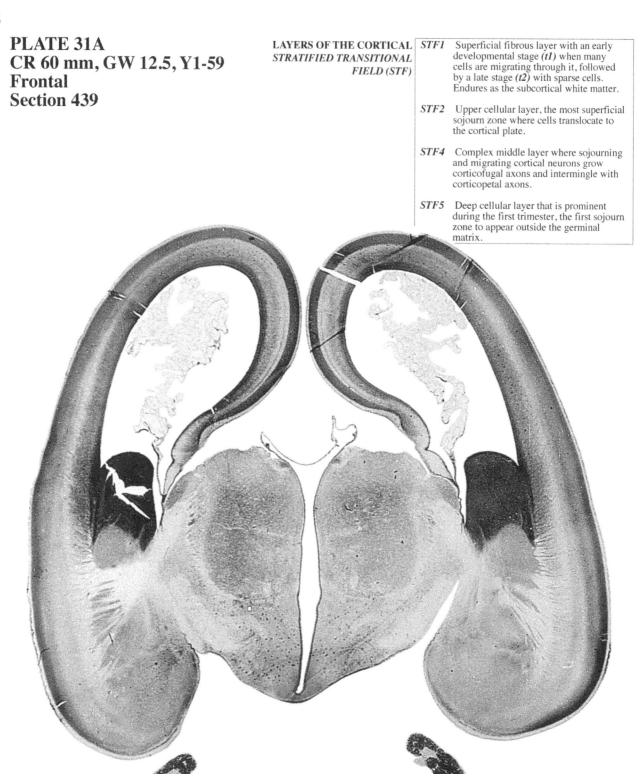

2 mm

See a high-magnification view of the diencephalon
from Section 449 in Plates 49A and B.

FONT KEY:
VENTRICULAR DIVISIONS – CAPITALS
Germinal zone - Helvetica bold
Transient structure - Times bold italic
Permanent structure - Times Roman or **Bold**

ABBREVIATIONS:
GEP - Glioepithelium
G/EP - Glioepithelium/ependyma
NEP - Neuroepithelium
SVZ - Subventricular zone

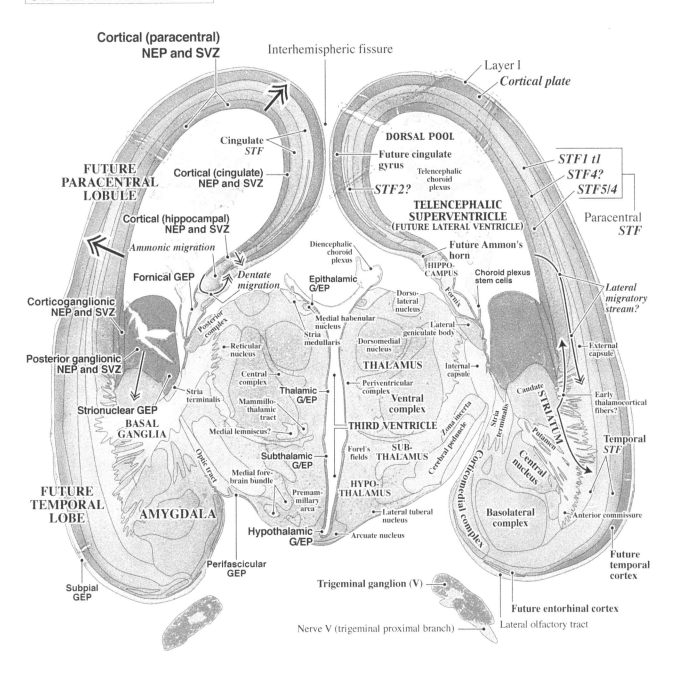

Cortical (paracentral) NEP and SVZ

Interhemispheric fissure

Layer I
Cortical plate

Cingulate *STF*

Cortical (cingulate) NEP and SVZ

FUTURE PARACENTRAL LOBULE

Cortical (hippocampal) NEP and SVZ

Ammonic migration

Fornical GEP

Corticoganglionic NEP and SVZ

Posterior ganglionic NEP and SVZ

Strionuclear GEP

BASAL GANGLIA

FUTURE TEMPORAL LOBE

AMYGDALA

Subpial GEP

Perifascicular GEP

DORSAL POOL

Future cingulate gyrus

Telencephalic choroid plexus

STF2?

TELENCEPHALIC SUPERVENTRICLE (FUTURE LATERAL VENTRICLE)

STF1 t1
STF4?
STF5/4

Paracentral *STF*

Diencephalic choroid plexus

Future Ammon's horn

HIPPO-CAMPUS

Choroid plexus stem cells

Lateral migratory stream?

Epithalamic G/EP

Dentate migration

Dorso-lateral nucleus

Fornix

External capsule

Posterior complex

Medial habenular nucleus

Stria medullaris

Lateral geniculate body

Reticular nucleus

Dorsomedial nucleus

THALAMUS

Internal capsule

Early thalamocortical fibers?

Central complex

Periventricular complex

Caudate

STRIATUM

Stria terminalis

Mammillo-thalamic tract

Thalamic G/EP

Ventral complex

Zona incerta

Stria terminalis

Putamen

Medial lemniscus?

THIRD VENTRICLE

Temporal *STF*

Subthalamic G/EP

Forel's fields

SUB-THALAMUS

Cerebral peduncle

Corticomedial complex

Central nucleus

Optic tract

Medial fore-brain bundle

Premammillary area

HYPO-THALAMUS

Basolateral complex

Anterior commissure

Hypothalamic G/EP

Lateral tuberal nucleus

Arcuate nucleus

Future temporal cortex

Trigeminal ganglion (V)

Future entorhinal cortex

Nerve V (trigeminal proximal branch)

Lateral olfactory tract

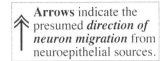

Arrows indicate the presumed *direction of axon growth* in brain fiber tracts.

Arrows indicate the presumed *direction of neuron migration* from neuroepithelial sources.

Dashed lines indicate staining and/or sectioning artifacts.

PLATE 32A
CR 60 mm, GW 12.5, Y1-59
Frontal
Section 459

LAYERS OF THE CORTICAL
STRATIFIED TRANSITIONAL
FIELD (STF)

STF1	Superficial fibrous layer with an early developmental stage *(t1)* when many cells are migrating through it, followed by a late stage *(t2)* with sparse cells. Endures as the subcortical white matter.
STF2	Upper cellular layer, the most superficial sojourn zone where cells translocate to the cortical plate.
STF4	Complex middle layer where sojourning and migrating cortical neurons grow corticofugal axons and intermingle with corticopetal axons.
STF5	Deep cellular layer that is prominent during the first trimester, the first sojourn zone to appear outside the germinal matrix.

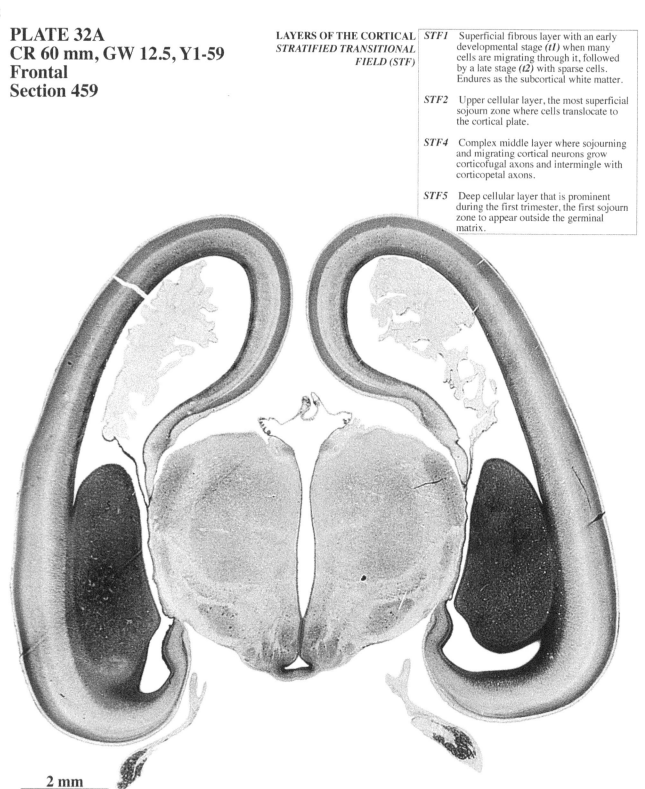

2 mm

See a high-magnification view of the diencephalon from Section 449 in Plates 49A and B, from Section 469 in Plates 50A and B.

FONT KEY:
VENTRICULAR DIVISIONS - CAPITALS
Germinal zone - Helvetica bold
Transient structure - Times bold italic
Permanent structure - Times Roman or **Bold**

ABBREVIATIONS:
GEP - Glioepithelium
G/EP - Glioepithelium/ependyma
NEP - Neuroepithelium
SVZ - Subventricular zone

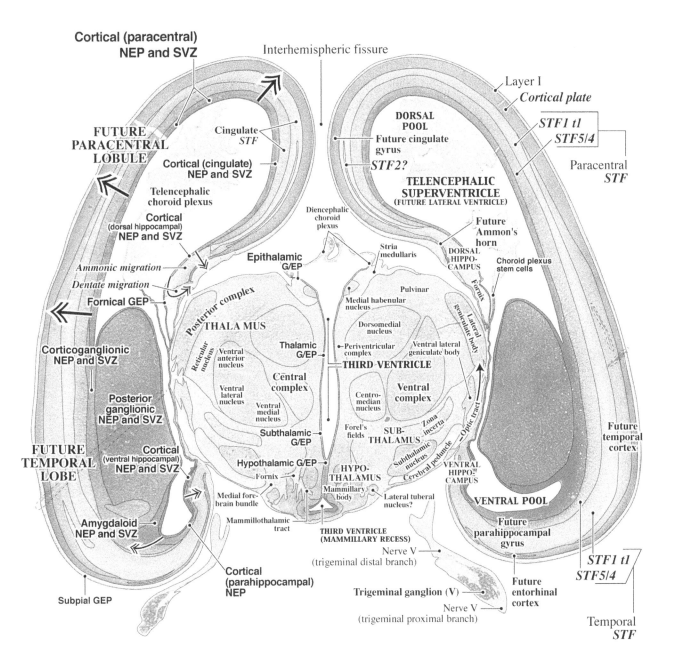

Cortical (paracentral)
NEP and SVZ

Interhemispheric fissure

Layer I
Cortical plate

FUTURE
PARACENTRAL
LOBULE

Cingulate
STF

DORSAL
POOL
Future cingulate
gyrus
STF2?

STF1 t1
STF5/4

Paracentral
STF

Cortical (cingulate)
NEP and SVZ

Telencephalic
choroid plexus

TELENCEPHALIC
SUPERVENTRICLE
(FUTURE LATERAL VENTRICLE)

Cortical
(dorsal hippocampal)
NEP and SVZ

Diencephalic
choroid
plexus

Future
Ammon's
horn

DORSAL
HIPPO-
CAMPUS

Choroid plexus
stem cells

Ammonic migration

Stria
medullaris

Epithalamic
G/EP

Dentate migration

Pulvinar

Medial habenular
nucleus

Fornical GEP

Posterior complex

THALAMUS

Dorsomedial
nucleus

Ventral lateral
geniculate body

Thalamic
G/EP

Periventricular
complex

Lateral
geniculate body

Corticoganglionic
NEP and SVZ

Reticular nucleus

Ventral
anterior
nucleus

THIRD VENTRICLE

Ventral
complex

Posterior
ganglionic
NEP and SVZ

Central
complex

Ventral
lateral
nucleus

Centro-
median
nucleus

Zona incerta

Ventral
medial
nucleus

FUTURE
TEMPORAL
LOBE

Cortical
(ventral hippocampal)
NEP and SVZ

Subthalamic
G/EP

Forel's
fields

SUB-
THALAMUS

Subthalamic
nucleus

Cerebral peduncle

Optic tract

VENTRAL
HIPPO-
CAMPUS

Future
temporal
cortex

Hypothalamic G/EP

HYPO-
THALAMUS

VENTRAL POOL

Fornix

Mammillary
body

Lateral tuberal
nucleus?

Amygdaloid
NEP and SVZ

Medial fore-
brain bundle

Mammillothalamic
tract

THIRD VENTRICLE
(MAMMILLARY RECESS)

Future
parahippocampal
gyrus

Nerve V
(trigeminal distal branch)

STF1 t1
STF5/4

Cortical
(parahippocampal)
NEP

Trigeminal ganglion (V)

Nerve V
(trigeminal proximal branch)

Future
entorhinal
cortex

Temporal
STF

Subpial GEP

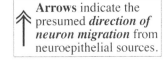

Arrows indicate the
presumed *direction of*
axon growth in brain
fiber tracts.

Arrows indicate the
presumed *direction of*
neuron migration from
neuroepithelial sources.

Dashed lines indicate staining
and/or sectioning artifacts.

PLATE 33A
CR 60 mm, GW 12.5, Y1-59
Frontal
Section 479

LAYERS OF THE CORTICAL
STRATIFIED TRANSITIONAL
FIELD (STF)

STF1 Superficial fibrous layer with an early developmental stage *(t1)* when many cells are migrating through it, followed by a late stage *(t2)* with sparse cells. Endures as the subcortical white matter.

STF2 Upper cellular layer, the most superficial sojourn zone where cells translocate to the cortical plate.

STF4 Complex middle layer where sojourning and migrating cortical neurons grow corticofugal axons and intermingle with corticopetal axons.

STF5 Deep cellular layer that is prominent during the first trimester, the first sojourn zone to appear outside the germinal matrix.

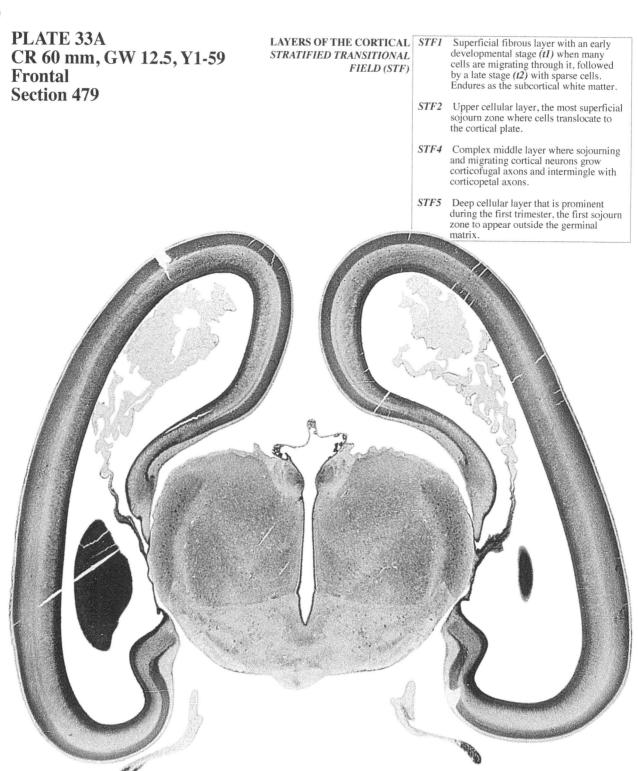

2 mm

See a high-magnification view of the diencephalon from Section 469 in Plates 50A and B.

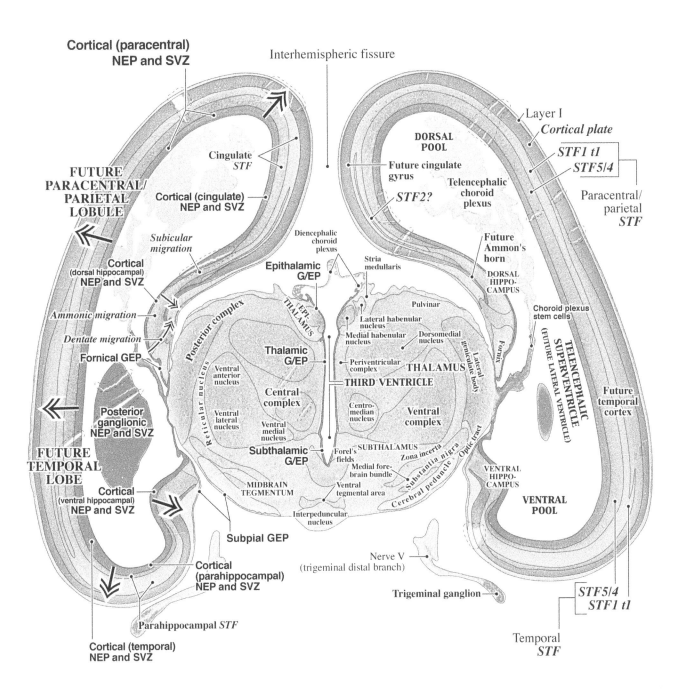

Arrows indicate the presumed *direction of neuron migration* from neuroepithelial sources.

Dashed lines indicate staining and/or sectioning artifacts.

PLATE 34A
CR 60 mm, GW 12.5, Y1-59
Frontal
Section 500

LAYERS OF THE CORTICAL
STRATIFIED TRANSITIONAL FIELD (STF)

STF1   Superficial fibrous layer with an early developmental stage *(t1)* when many cells are migrating through it, followed by a late stage *(t2)* with sparse cells. Endures as the subcortical white matter.

STF2   Upper cellular layer, the most superficial sojourn zone where cells translocate to the cortical plate.

STF4   Complex middle layer where sojourning and migrating cortical neurons grow corticofugal axons and intermingle with corticopetal axons.

STF5   Deep cellular layer that is prominent during the first trimester, the first sojourn zone to appear outside the germinal matrix.

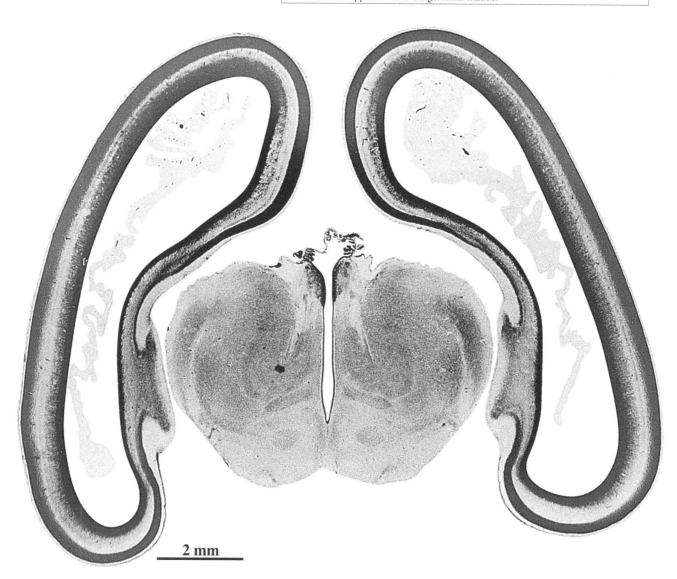

2 mm

See a high-magnification view of the midbrain and thalamus from Section 499 in Plates 51A and B.

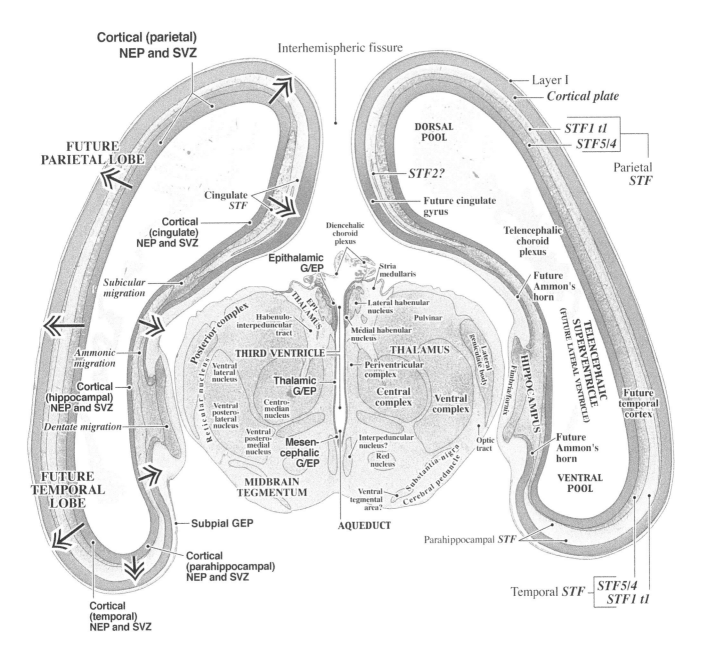

Arrows indicate the presumed *direction of neuron migration* from neuroepithelial sources.

Dashed lines indicate staining and/or sectioning artifacts.

PLATE 35A
CR 60 mm, GW 12.5, Y1-59
Frontal
Section 529

LAYERS OF THE CORTICAL
STRATIFIED TRANSITIONAL FIELD (STF)

STF1 Superficial fibrous layer with an early developmental stage *(t1)* when many cells are migrating through it, followed by a late stage *(t2)* with sparse cells. Endures as the subcortical white matter.

STF2 Upper cellular layer, the most superficial sojourn zone where cells translocate to the cortical plate.

STF4 Complex middle layer where sojourning and migrating cortical neurons grow corticofugal axons and intermingle with corticopetal axons.

STF5 Deep cellular layer that is prominent during the first trimester, the first sojourn zone to appear outside the germinal matrix.

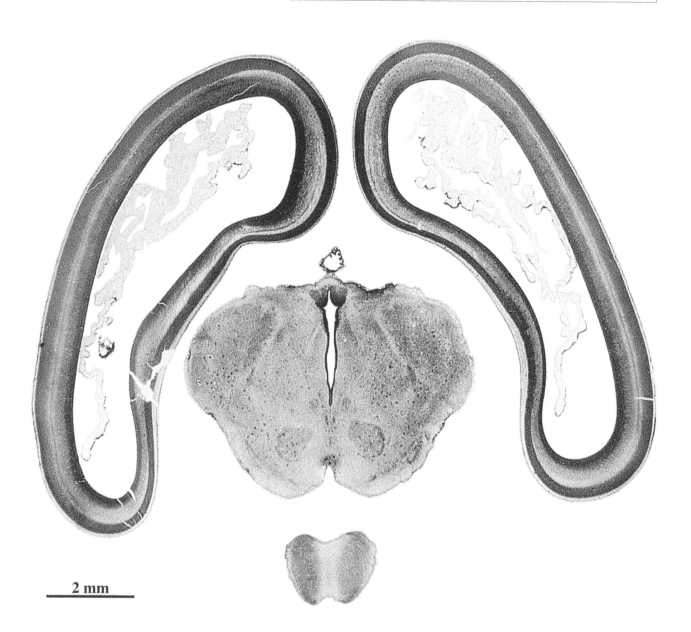

2 mm

See a high-magnification view of the parietal cortex from Section 519 in Plates 44A and B, and of the thalamus and midbrain from this Section in Plates 52A and B.

FONT KEY:
VENTRICULAR DIVISIONS - CAPITALS
Germinal zone - Helvetica bold
Transient structure - Times bold italic
Permanent structure - Times Roman or **Bold**

ABBREVIATIONS:
GEP - Glioepithelium
G/EP - Glioepithelium/ependyma
NEP - Neuroepithelium
SVZ - Subventricular zone

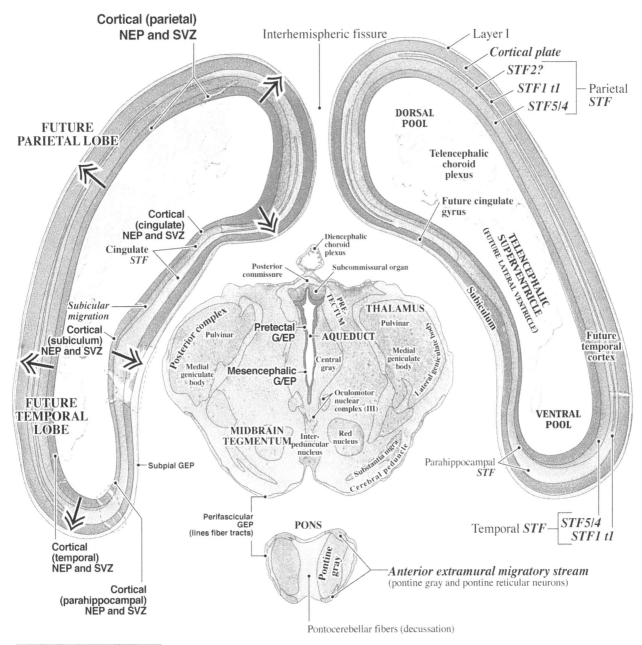

Cortical (parietal)
NEP and SVZ

Interhemispheric fissure

Layer I

Cortical plate

STF2?

STF1 t1 — Parietal *STF*

STF5/4

DORSAL POOL

Telencephalic choroid plexus

Future cingulate gyrus

FUTURE PARIETAL LOBE

Cortical (cingulate) NEP and SVZ

Cingulate *STF*

TELENCEPHALIC SUPERVENTRICLE (FUTURE LATERAL VENTRICLE)

Subiculum

Subicular migration

Cortical (subiculum) NEP and SVZ

Diencephalic choroid plexus

Posterior commissure

Subcommissural organ

PRE-TECTUM

THALAMUS

Pulvinar

Future temporal cortex

Posterior complex

Pulvinar

Pretectal G/EP

AQUEDUCT

Central gray

Medial geniculate body

Lateral geniculate body

Medial geniculate body

Mesencephalic G/EP

FUTURE TEMPORAL LOBE

Oculomotor nuclear complex (III)

VENTRAL POOL

MIDBRAIN TEGMENTUM

Inter-peduncular nucleus

Red nucleus

Substantia nigra

Cerebral peduncle

Parahippocampal *STF*

Subpial GEP

Perifascicular GEP (lines fiber tracts)

PONS

Pontine gray

Temporal *STF* — *STF5/4* / *STF1 t1*

Cortical (temporal) NEP and SVZ

Anterior extramural migratory stream
(pontine gray and pontine reticular neurons)

Cortical (parahippocampal) NEP and SVZ

Pontocerebellar fibers (decussation)

Arrows indicate the presumed *direction of neuron migration* from neuroepithelial sources.

Dashed lines indicate staining and/or sectioning artifacts.

86

**PLATE 36A
CR 60 mm, GW 12.5, Y1-59
Frontal
Section 569**

LAYERS OF THE CORTICAL
STRATIFIED TRANSITIONAL FIELD (STF)

STF1 Superficial fibrous layer with an early developmental stage *(t1)* when many cells are migrating through it, followed by a late stage *(t2)* with sparse cells. Endures as the subcortical white matter.

STF2 Upper cellular layer, the most superficial sojourn zone where cells translocate to the cortical plate.

STF4 Complex middle layer where sojourning and migrating cortical neurons grow corticofugal axons and intermingle with corticopetal axons.

STF5 Deep cellular layer that is prominent during the first trimester, the first sojourn zone to appear outside the germinal matrix.

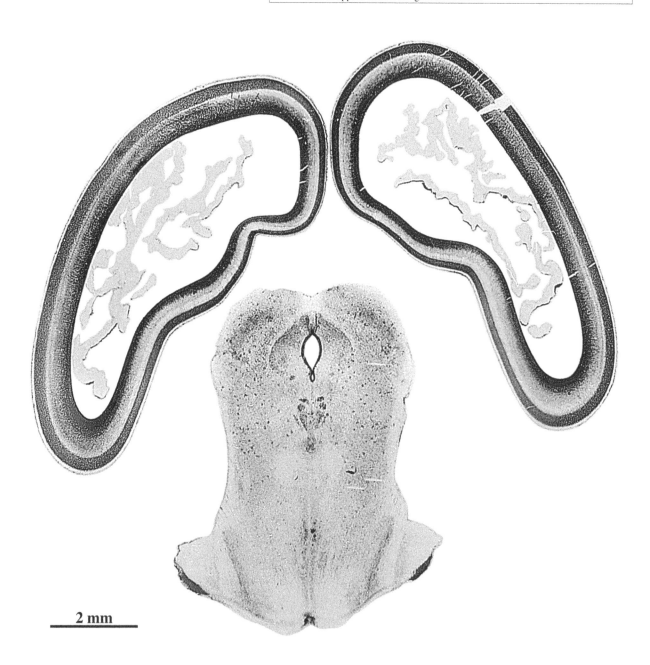

2 mm

See a high-magnification view of the midbrain and pons from Section 549 in Plates 32A and B, and from this Section in Plates 54A and B.

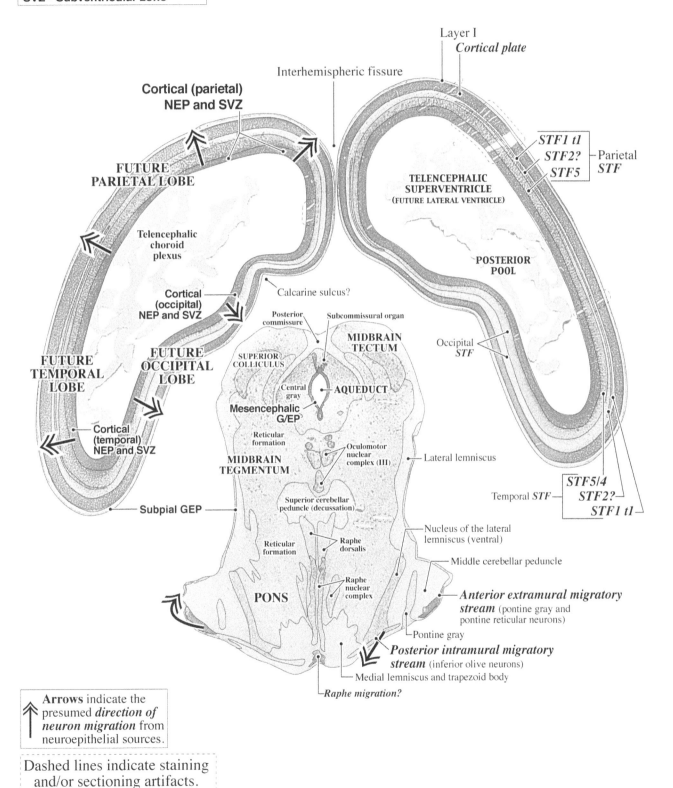

FONT KEY:
VENTRICULAR DIVISIONS - CAPITALS
Germinal zone - Helvetica bold
Transient structure - Times bold italic
Permanent structure - Times Roman or **Bold**

ABBREVIATIONS:
GEP - Glioepithelium
G/EP - Glioepithelium/ependyma
NEP - Neuroepithelium
SVZ - Subventricular zone

Layer I
Cortical plate

Interhemispheric fissure

**Cortical (parietal)
NEP and SVZ**

STF1 t1
STF2?
STF5
Parietal
STF

**FUTURE
PARIETAL LOBE**

**TELENCEPHALIC
SUPERVENTRICLE**
(FUTURE LATERAL VENTRICLE)

Telencephalic
choroid
plexus

**POSTERIOR
POOL**

**Cortical
(occipital)
NEP and SVZ**

Calcarine sulcus?

Posterior
commissure

Subcommissural organ

Occipital
STF

**MIDBRAIN
TECTUM**

**FUTURE
TEMPORAL
LOBE**

**FUTURE
OCCIPITAL
LOBE**

**SUPERIOR
COLLICULUS**

Central
gray

AQUEDUCT

**Mesencephalic
G/EP**

Reticular
formation

Oculomotor
nuclear
complex (III)

Lateral lemniscus

**Cortical
(temporal)
NEP and SVZ**

**MIDBRAIN
TEGMENTUM**

STF5/4
STF2?
STF1 t1

Temporal *STF*

Superior cerebellar
peduncle (decussation)

Nucleus of the lateral
lemniscus (ventral)

Subpial GEP

Reticular
formation

Raphe
dorsalis

Middle cerebellar peduncle

Raphe
nuclear
complex

PONS

*Anterior extramural migratory
stream* (pontine gray and
pontine reticular neurons)

Pontine gray

*Posterior intramural migratory
stream* (inferior olive neurons)

Medial lemniscus and trapezoid body

Raphe migration?

Arrows indicate the
presumed *direction of
neuron migration* from
neuroepithelial sources.

Dashed lines indicate staining
and/or sectioning artifacts.

PLATE 37A
CR 60 mm, GW 12.5, Y1-59
Frontal
Section 599

LAYERS OF THE CORTICAL
STRATIFIED TRANSITIONAL FIELD (STF)

STF1 Superficial fibrous layer with an early developmental stage *(t1)* when many cells are migrating through it, followed by a late stage *(t2)* with sparse cells. Endures as the subcortical white matter.

STF4 Complex middle layer where sojourning and migrating cortical neurons grow corticofugal axons and intermingle with corticopetal axons.

STF5 Deep cellular layer that is prominent during the first trimester, the first sojourn zone to appear outside the germinal matrix.

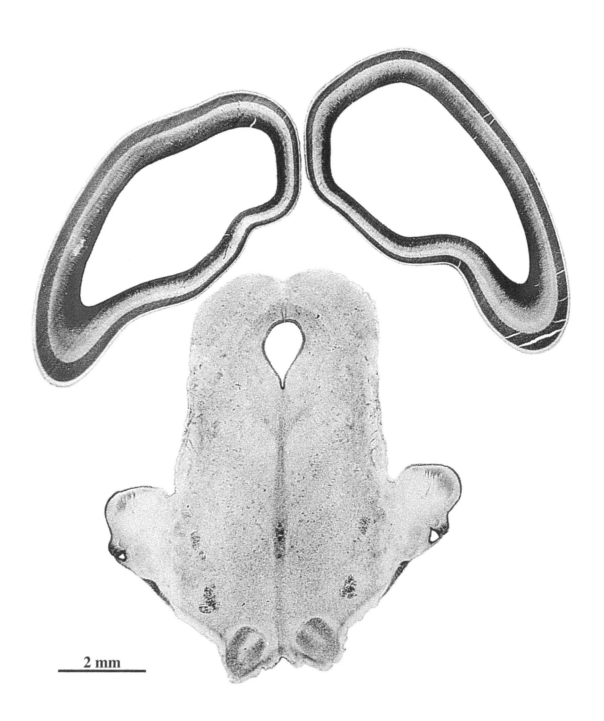

2 mm

See a high-magnification view of the midbrain and pons from section 589 in Plates 55A and B and from this Section in Plates 56A and B.

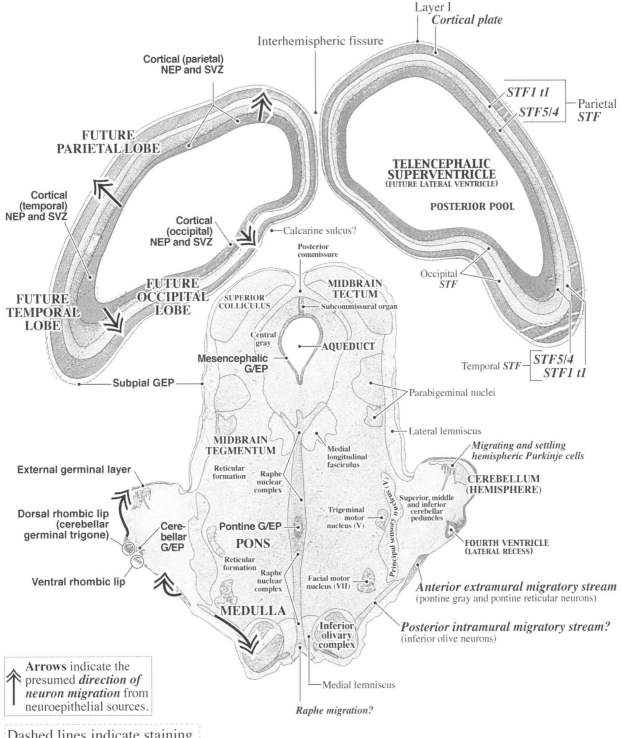

Interhemispheric fissure

Layer I
Cortical plate

Cortical (parietal)
NEP and SVZ

STF1 tl
STF5/4
Parietal
STF

FUTURE
PARIETAL LOBE

TELENCEPHALIC
SUPERVENTRICLE
(FUTURE LATERAL VENTRICLE)

POSTERIOR POOL

Cortical
(temporal)
NEP and SVZ

Cortical
(occipital)
NEP and SVZ

Calcarine sulcus?

Posterior
commissure

Occipital
STF

FUTURE
OCCIPITAL
LOBE

FUTURE
TEMPORAL
LOBE

SUPERIOR
COLLICULUS

MIDBRAIN
TECTUM

Subcommissural organ

Central
gray

Mesencephalic
G/EP

AQUEDUCT

Temporal *STF*

STF5/4
STF1 tl

Parabigeminal nuclei

Subpial GEP

Lateral lemniscus

MIDBRAIN
TEGMENTUM

Medial
longitudinal
fasciculus

*Migrating and settling
hemispheric Purkinje cells*

CEREBELLUM
(HEMISPHERE)

External germinal layer

Reticular
formation

Raphe
nuclear
complex

Superior, middle
and inferior
cerebellar
peduncles

Dorsal rhombic lip
(cerebellar
germinal trigone)

Cere-
bellar
G/EP

Pontine G/EP

PONS

Trigeminal
motor
nucleus (V)

Principal sensory nucleus (V)

Ventral rhombic lip

FOURTH VENTRICLE
(LATERAL RECESS)

Reticular
formation

Raphe
nuclear
complex

Facial motor
nucleus (VII)

Anterior extramural migratory stream
(pontine gray and pontine reticular neurons)

MEDULLA

Inferior
olivary
complex

Posterior intramural migratory stream?
(inferior olive neurons)

Medial lemniscus

Raphe migration?

Arrows indicate the
presumed *direction of
neuron migration* from
neuroepithelial sources.

Dashed lines indicate staining
and/or sectioning artifacts.

PLATE 38A
CR 60 mm, GW 12.5, Y1-59
Frontal
Section 629

LAYERS OF THE CORTICAL
STRATIFIED TRANSITIONAL FIELD (STF)

STF1 Superficial fibrous layer with an early developmental stage *(t1)* when many cells are migrating through it, followed by a late stage *(t2)* with sparse cells. Endures as the subcortical white matter.

STF4 Complex middle layer where sojourning and migrating cortical neurons grow corticofugal axons and intermingle with corticopetal axons.

STF5 Deep cellular layer that is prominent during the first trimester, the first sojourn zone to appear outside the germinal matrix.

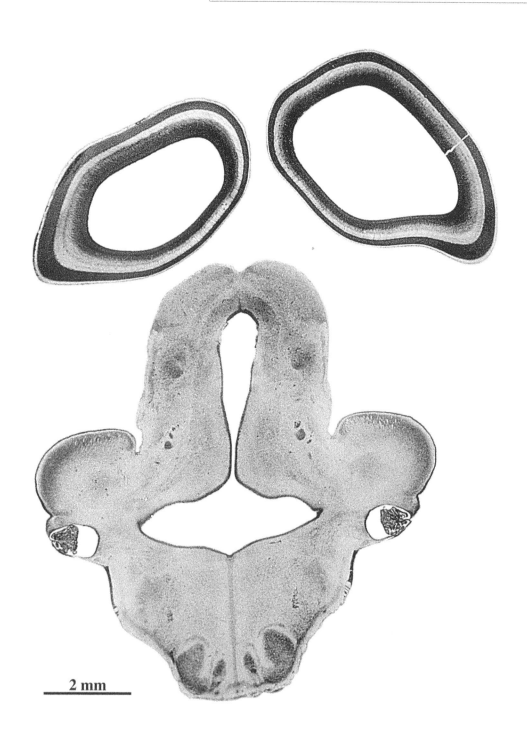

2 mm

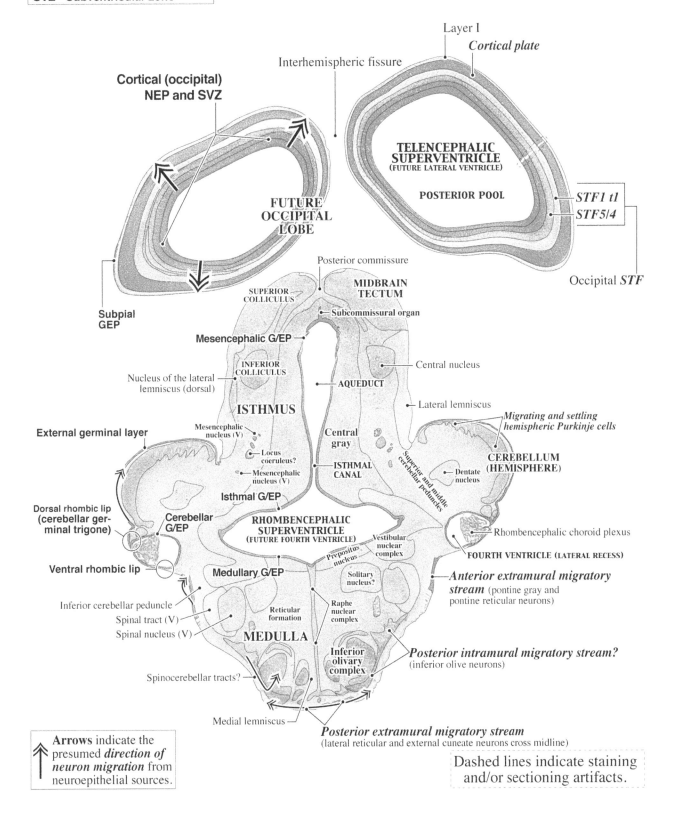

Layer I
Cortical plate

Interhemispheric fissure

Cortical (occipital)
NEP and SVZ

**TELENCEPHALIC
SUPERVENTRICLE**
(FUTURE LATERAL VENTRICLE)

POSTERIOR POOL

STF1 t1
STF5/4

**FUTURE
OCCIPITAL
LOBE**

Occipital *STF*

Subpial
GEP

Posterior commissure

**MIDBRAIN
TECTUM**

SUPERIOR
COLLICULUS

Subcommissural organ

Mesencephalic G/EP

Central nucleus

INFERIOR
COLLICULUS

Nucleus of the lateral
lemniscus (dorsal)

AQUEDUCT

Lateral lemniscus

*Migrating and settling
hemispheric Purkinje cells*

ISTHMUS

Central
gray

Mesencephalic
nucleus (V)

**CEREBELLUM
(HEMISPHERE)**

External germinal layer

Locus
coeruleus?

**ISTHMAL
CANAL**

Superior and middle
cerebellar peduncles

Dentate
nucleus

Mesencephalic
nucleus (V)

Isthmal G/EP

Dorsal rhombic lip
(cerebellar ger-
minal trigone)

Cerebellar
G/EP

**RHOMBENCEPHALIC
SUPERVENTRICLE**
(FUTURE FOURTH VENTRICLE)

Vestibular
nuclear
complex

Rhombencephalic choroid plexus

Ventral rhombic lip

Medullary G/EP

Prepositus
nucleus

FOURTH VENTRICLE (LATERAL RECESS)

*Anterior extramural migratory
stream* (pontine gray and
pontine reticular neurons)

Inferior cerebellar peduncle

Solitary
nucleus?

Spinal tract (V)

Reticular
formation

Raphe
nuclear
complex

Spinal nucleus (V)

MEDULLA

Posterior intramural migratory stream?
(inferior olive neurons)

Inferior
olivary
complex

Spinocerebellar tracts?

Medial lemniscus

Posterior extramural migratory stream
(lateral reticular and external cuneate neurons cross midline)

Arrows indicate the
presumed *direction of
neuron migration* from
neuroepithelial sources.

Dashed lines indicate staining
and/or sectioning artifacts.

PLATE 39A
CR 60 mm, GW 12.5, Y1-59
Frontal
Section 649

LAYERS OF THE CORTICAL
STRATIFIED TRANSITIONAL FIELD (STF)

STF1 Superficial fibrous layer with an early developmental stage *(t1)* when many cells are migrating through it, followed by a late stage *(t2)* with sparse cells. Endures as the subcortical white matter.

STF4 Complex middle layer where sojourning and migrating cortical neurons grow corticofugal axons and intermingle with corticopetal axons.

STF5 Deep cellular layer that is prominent during the first trimester, the first sojourn zone to appear outside the germinal matrix.

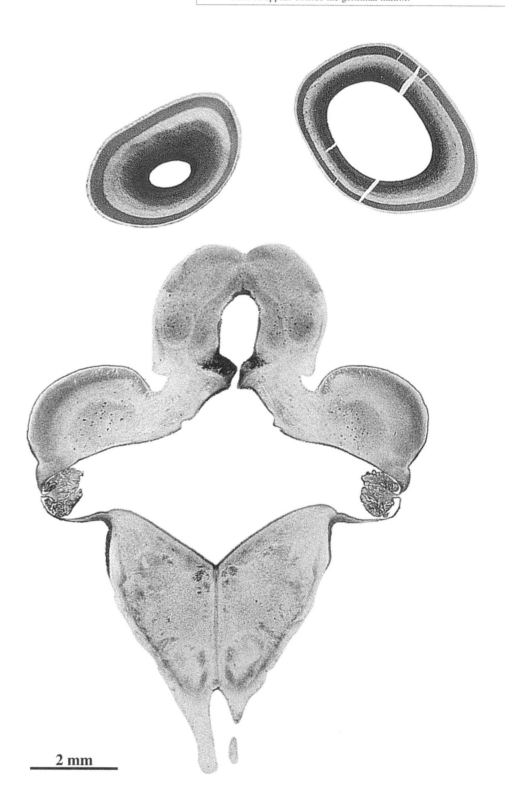

2 mm

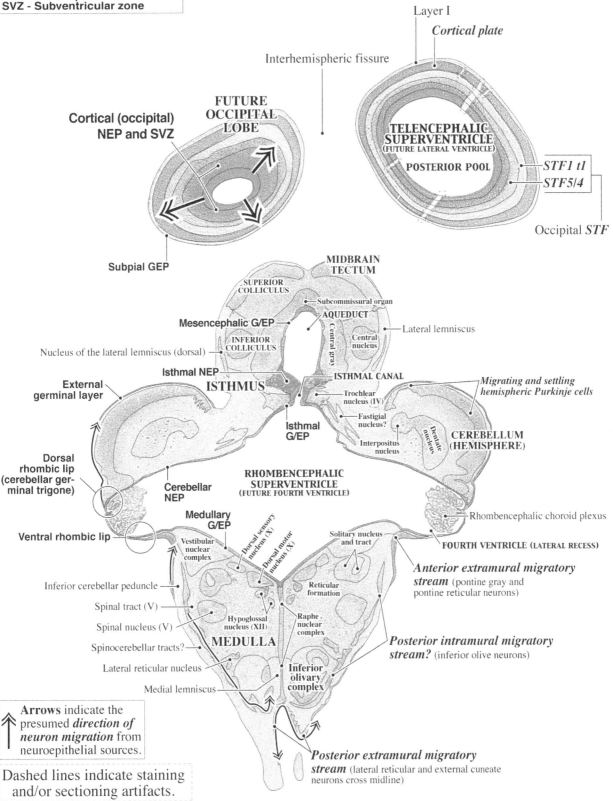

FONT KEY:
VENTRICULAR DIVISIONS - CAPITALS
Germinal zone - Helvetica bold
Transient structure - Times bold italic
Permanent structure - Times Roman or **Bold**

ABBREVIATIONS:
GEP - Glioepithelium
G/EP - Glioepithelium/ependyma
NEP - Neuroepithelium
SVZ - Subventricular zone

Layer I
Cortical plate

Interhemispheric fissure

**FUTURE
OCCIPITAL
LOBE**

Cortical (occipital)
NEP and SVZ

**TELENCEPHALIC
SUPERVENTRICLE**
(FUTURE LATERAL VENTRICLE)

POSTERIOR POOL

STF1 t1
STF5/4

Occipital *STF*

Subpial GEP

**MIDBRAIN
TECTUM**

**SUPERIOR
COLLICULUS**

Subcommissural organ

AQUEDUCT

Mesencephalic G/EP

Central gray

Central
nucleus

Lateral lemniscus

Nucleus of the lateral lemniscus (dorsal)

**INFERIOR
COLLICULUS**

Isthmal NEP

ISTHMUS

ISTHMAL CANAL

*Migrating and settling
hemispheric Purkinje cells*

External
germinal layer

Trochlear
nucleus (IV)

Fastigial
nucleus?

Dentate
nucleus

**CEREBELLUM
(HEMISPHERE)**

Isthmal
G/EP

Interpositus
nucleus

Dorsal
rhombic lip
(cerebellar ger-
minal trigone)

Cerebellar
NEP

**RHOMBENCEPHALIC
SUPERVENTRICLE**
(FUTURE FOURTH VENTRICLE)

Rhombencephalic choroid plexus

Ventral rhombic lip

Medullary
G/EP

Vestibular
nuclear
complex

Dorsal sensory
nucleus (X)

Dorsal motor
nucleus (X)

Solitary nucleus
and tract

FOURTH VENTRICLE (LATERAL RECESS)

Inferior cerebellar peduncle

Reticular
formation

*Anterior extramural migratory
stream* (pontine gray and
pontine reticular neurons)

Spinal tract (V)

Spinal nucleus (V)

Hypoglossal
nucleus (XII)

Raphe
nuclear
complex

MEDULLA

*Posterior intramural migratory
stream?* (inferior olive neurons)

Spinocerebellar tracts?

Lateral reticular nucleus

**Inferior
olivary
complex**

Medial lemniscus

Arrows indicate the
presumed *direction of
neuron migration* from
neuroepithelial sources.

*Posterior extramural migratory
stream* (lateral reticular and external cuneate
neurons cross midline)

Dashed lines indicate staining
and/or sectioning artifacts.

PLATE 40A
CR 60 mm, GW 12.5, Y1-59
Frontal
Section 659

LAYERS OF THE CORTICAL
STRATIFIED TRANSITIONAL FIELD (STF)

STF1 Superficial fibrous layer with an early developmental stage *(t1)* when many cells are migrating through it, followed by a late stage *(t2)* with sparse cells. Endures as the subcortical white matter.

STF4 Complex middle layer where sojourning and migrating cortical neurons grow corticofugal axons and intermingle with corticopetal axons.

STF5 Deep cellular layer that is prominent during the first trimester, the first sojourn zone to appear outside the germinal matrix.

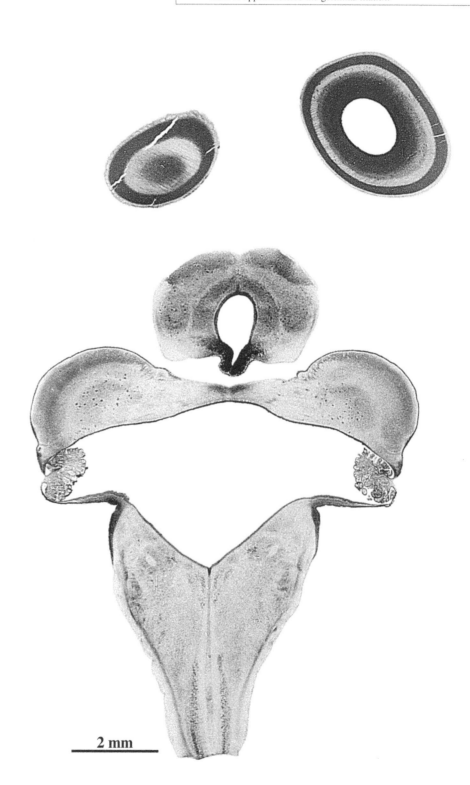

2 mm

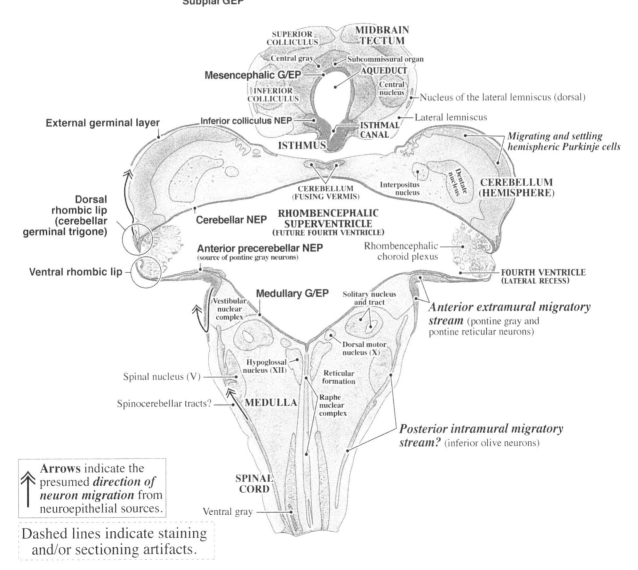

FONT KEY:
VENTRICULAR DIVISIONS - CAPITALS
Germinal zone - Helvetica bold
Transient structure - Times bold italic
Permanent structure - Times Roman or **Bold**

ABBREVIATIONS:
GEP - Glioepithelium
G/EP - Glioepithelium/ependyma
NEP - Neuroepithelium
SVZ - Subventricular zone

Interhemispheric fissure

Layer I

Cortical plate

FUTURE OCCIPITAL LOBE

STF1 t1

STF5/4

Subpial GEP

TELENCEPHALIC SUPERVENTRICLE
(FUTURE LATERAL VENTRICLE, POSTERIOR POOL)

STF1 t1
STF5/4

Occipital *STF*

Cortical (occipital) NEP and SVZ

SUPERIOR COLLICULUS

MIDBRAIN TECTUM

Central gray

Subcommissural organ

Mesencephalic G/EP

AQUEDUCT

Central nucleus

INFERIOR COLLICULUS

Nucleus of the lateral lemniscus (dorsal)

External germinal layer

Inferior colliculus NEP

Lateral lemniscus

ISTHMAL CANAL

ISTHMUS

Migrating and settling hemispheric Purkinje cells

Dorsal rhombic lip (cerebellar germinal trigone)

CEREBELLUM (FUSING VERMIS)

Interpositus nucleus

Dentate nucleus

CEREBELLUM (HEMISPHERE)

Cerebellar NEP

RHOMBENCEPHALIC SUPERVENTRICLE
(FUTURE FOURTH VENTRICLE)

Anterior precerebellar NEP
(source of pontine gray neurons)

Rhombencephalic choroid plexus

Ventral rhombic lip

FOURTH VENTRICLE
(LATERAL RECESS)

Vestibular nuclear complex

Medullary G/EP

Solitary nucleus and tract

Anterior extramural migratory stream (pontine gray and pontine reticular neurons)

Dorsal motor nucleus (X)

Hypoglossal nucleus (XII)

Reticular formation

Spinal nucleus (V)

MEDULLA

Raphe nuclear complex

Spinocerebellar tracts?

Posterior intramural migratory stream? (inferior olive neurons)

SPINAL CORD

Ventral gray

Arrows indicate the presumed *direction of neuron migration* from neuroepithelial sources.

Dashed lines indicate staining and/or sectioning artifacts.

PLATE 41A
CR 60 mm, GW 12.5, Y1-59
Frontal
Section 680

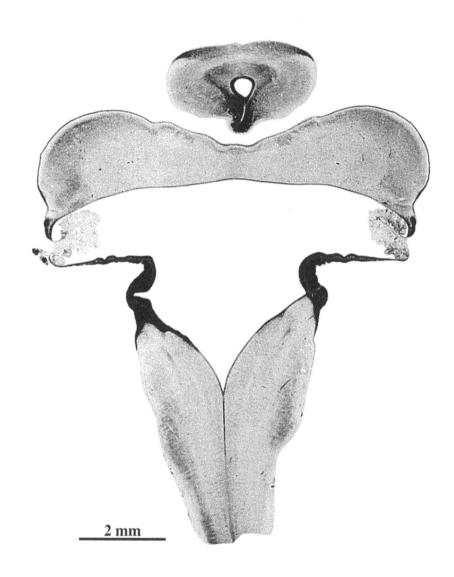

2 mm

FONT KEY:
VENTRICULAR DIVISIONS - CAPITALS
Germinal zone - Helvetica bold
Transient structure - Times bold italic
Permanent structure - Times Roman or **Bold**

ABBREVIATIONS:
G/EP - Glioepithelium/ependyma
NEP - Neuroepithelium

Arrows indicate the presumed *direction of neuron migration* from neuroepithelial sources.

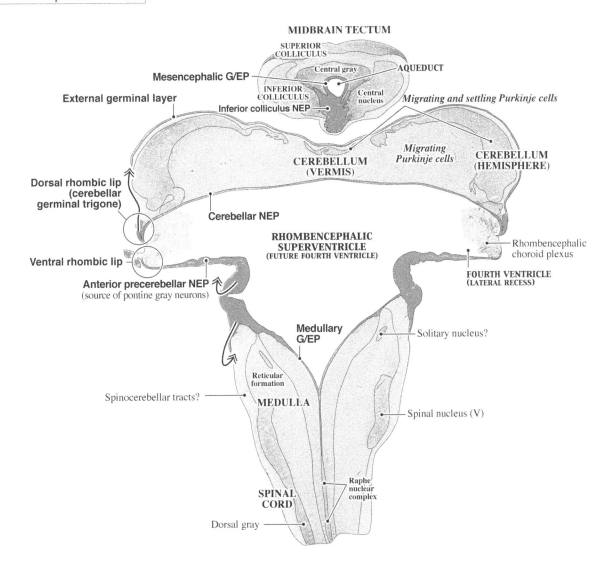

MIDBRAIN TECTUM

SUPERIOR COLLICULUS

Central gray

AQUEDUCT

Mesencephalic G/EP

INFERIOR COLLICULUS

Central nucleus

External germinal layer

Inferior colliculus NEP

Migrating and settling Purkinje cells

Migrating Purkinje cells

CEREBELLUM (VERMIS)

CEREBELLUM (HEMISPHERE)

Dorsal rhombic lip (cerebellar germinal trigone)

Cerebellar NEP

RHOMBENCEPHALIC SUPERVENTRICLE (FUTURE FOURTH VENTRICLE)

Rhombencephalic choroid plexus

Ventral rhombic lip

FOURTH VENTRICLE (LATERAL RECESS)

Anterior precerebellar NEP
(source of pontine gray neurons)

Medullary G/EP

Solitary nucleus?

Reticular formation

Spinocerebellar tracts?

MEDULLA

Spinal nucleus (V)

SPINAL CORD

Raphe nuclear complex

Dorsal gray

PLATE 42A
CR 60 mm, GW 12.5, Y1-59
Frontal
Y1-59

Section 709

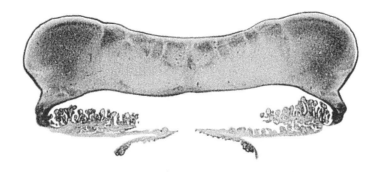

Section 720

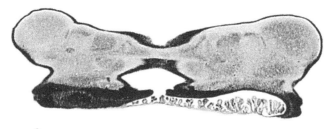

2 mm

Neuroepithelium - NEP

Section 709

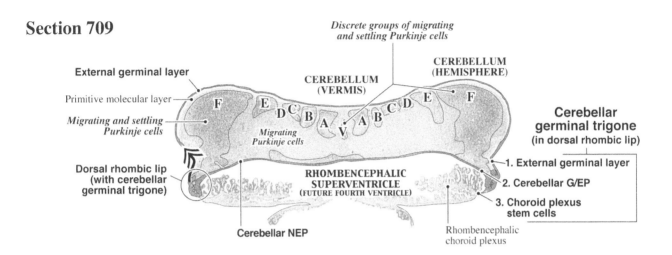

Discrete groups of migrating and settling Purkinje cells

CEREBELLUM (VERMIS)

CEREBELLUM (HEMISPHERE)

External germinal layer

Primitive molecular layer

Migrating and settling Purkinje cells

Migrating Purkinje cells

Dorsal rhombic lip (with cerebellar germinal trigone)

RHOMBENCEPHALIC SUPERVENTRICLE (FUTURE FOURTH VENTRICLE)

Cerebellar NEP

Rhombencephalic choroid plexus

Cerebellar germinal trigone (in dorsal rhombic lip)

1. **External germinal layer**
2. **Cerebellar G/EP**
3. **Choroid plexus stem cells**

Section 720

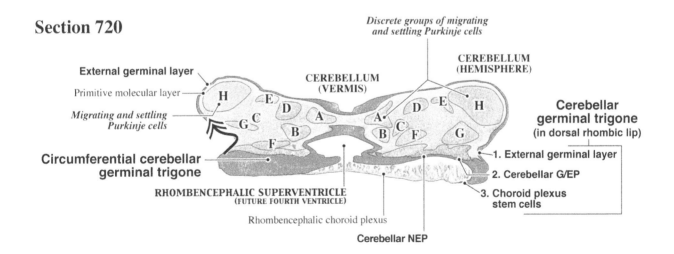

Discrete groups of migrating and settling Purkinje cells

CEREBELLUM (VERMIS)

CEREBELLUM (HEMISPHERE)

External germinal layer

Primitive molecular layer

Migrating and settling Purkinje cells

Circumferential cerebellar germinal trigone

RHOMBENCEPHALIC SUPERVENTRICLE (FUTURE FOURTH VENTRICLE)

Rhombencephalic choroid plexus

Cerebellar NEP

Cerebellar germinal trigone (in dorsal rhombic lip)

1. **External germinal layer**
2. **Cerebellar G/EP**
3. **Choroid plexus stem cells**

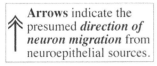

Arrows indicate the presumed *direction of neuron migration* from neuroepithelial sources.

PLATE 43A

CR 60 mm, GW 12.5, Y1-59
Frontal
Section 269
FRONTAL
CORTEX

0.25 mm

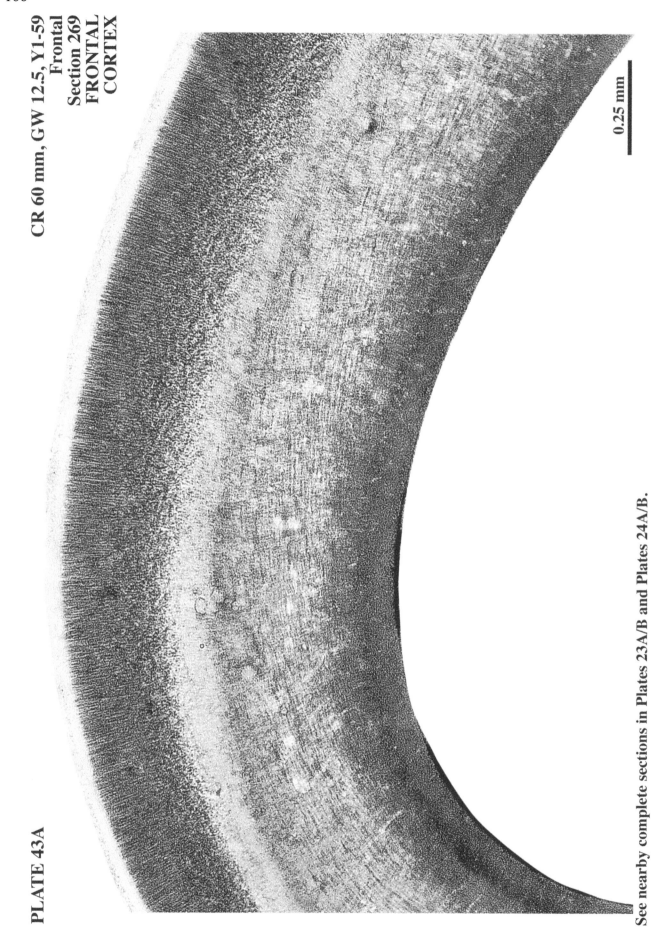

See nearby complete sections in Plates 23A/B and Plates 24A/B.

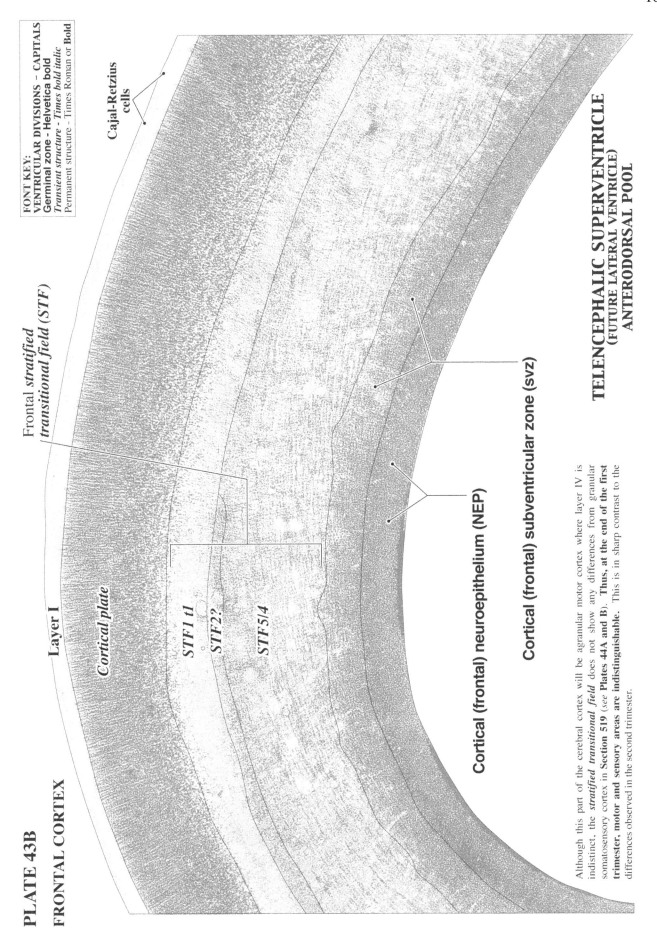

FONT KEY:
VENTRICULAR DIVISIONS – CAPITALS
Germinal zone - **Helvetica bold**
Transient structure - Times bold italic
Permanent structure - Times Roman or **Bold**

PLATE 43B
FRONTAL CORTEX

Cajal-Retzius cells

Frontal *stratified transitional field (STF)*

Layer I

Cortical plate

STF1 t1

STF2?

STF5/4

Cortical (frontal) neuroepithelium (NEP)

Cortical (frontal) subventricular zone (svz)

TELENCEPHALIC SUPERVENTRICLE
(FUTURE LATERAL VENTRICLE)
ANTERODORSAL POOL

Although this part of the cerebral cortex will be agranular motor cortex where layer IV is indistinct, the *stratified transitional field* does not show any differences from granular somatosensory cortex in **Section 519** (*see* **Plates 44A and B**). **Thus, at the end of the first trimester, motor and sensory areas are indistinguishable.** This is in sharp contrast to the differences observed in the second trimester.

PLATE 44A

CR 60 mm, GW 12.5, Y1-59
Frontal
Section 519
PARIETAL
CORTEX

0.25 mm

See the entire section in Plates 35A/B.

PLATE 44B
PARIETAL CORTEX

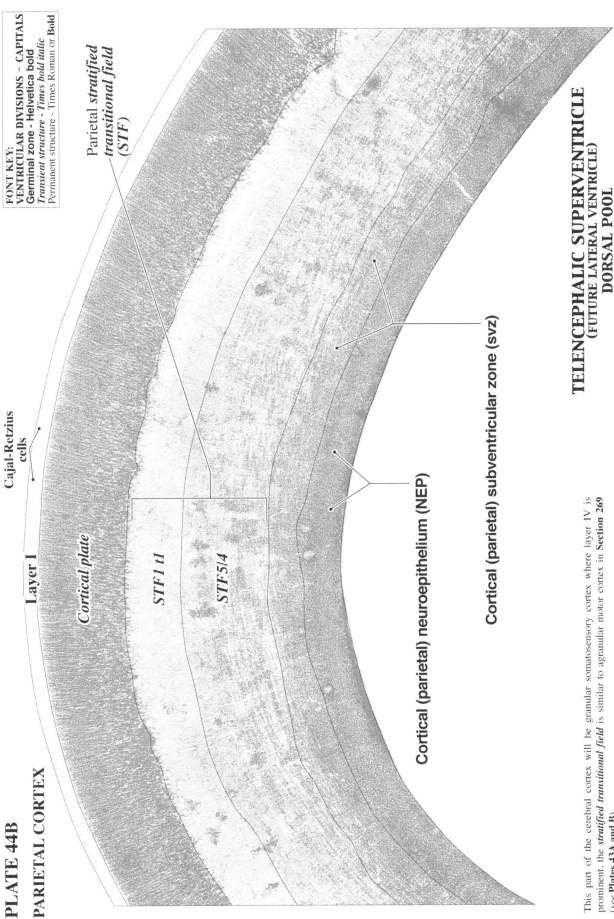

Cajal-Retzius
cells

Layer I

Cortical plate

*Parietal stratified
transitional field
(STF)*

STF1 t1

STF5/4

Cortical (parietal) neuroepithelium (NEP)

Cortical (parietal) subventricular zone (svz)

TELENCEPHALIC SUPERVENTRICLE
(FUTURE LATERAL VENTRICLE)
DORSAL POOL

This part of the cerebral cortex will be granular somatosensory cortex where layer IV is prominent, the *stratified transitional field* is similar to agranular motor cortex in **Section 269** (*see* **Plates 43A and B**).

PLATE 45A

**PARACENTRAL
CORTEX**

See nearby
sections in Plates
29 to 30A and B.

Enlarged in
Plates 46A and B.

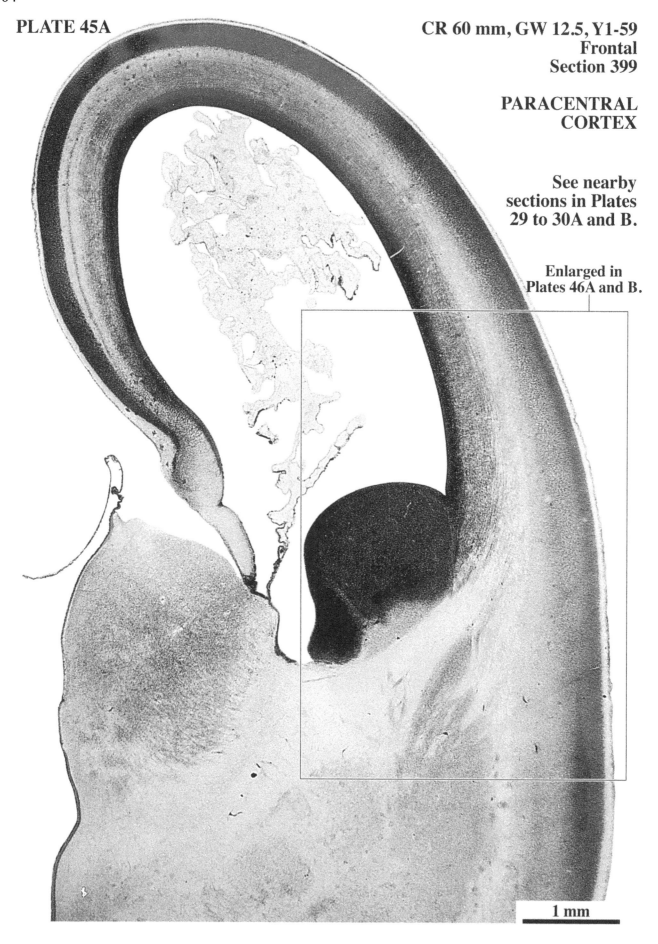

1 mm

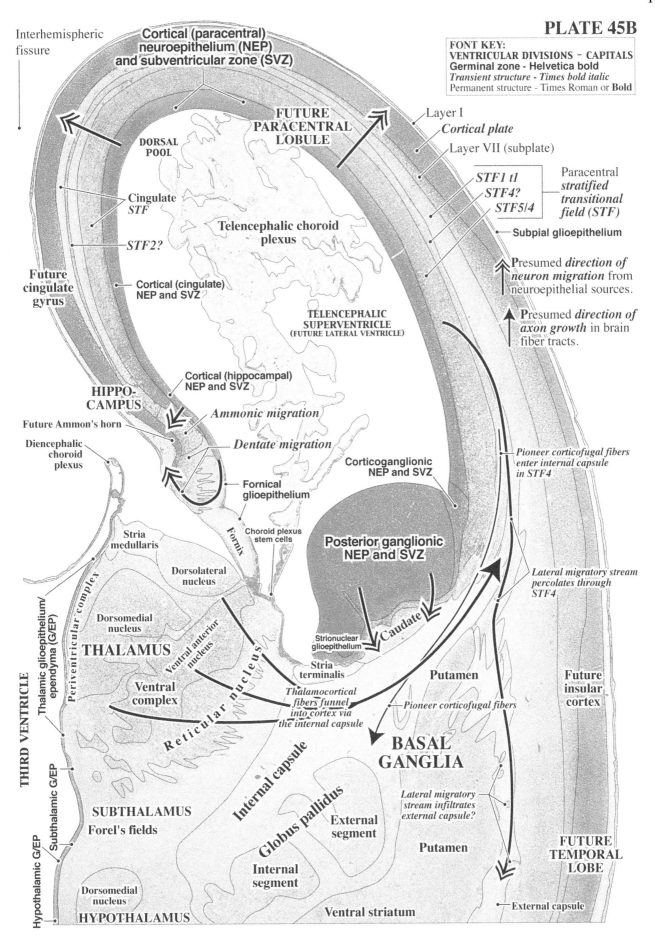

FONT KEY:
VENTRICULAR DIVISIONS – CAPITALS
Germinal zone - Helvetica bold
Transient structure - Times bold italic
Permanent structure - Times Roman or **Bold**

Interhemispheric fissure

Cortical (paracentral) neuroepithelium (NEP) and subventricular zone (SVZ)

FUTURE PARACENTRAL LOBULE

Layer I
Cortical plate
Layer VII (subplate)

STF1 t1
STF4?
STF5/4

Paracentral *stratified transitional field (STF)*

Subpial glioepithelium

DORSAL POOL

Cingulate *STF*

STF2?

Future cingulate gyrus

Cortical (cingulate) NEP and SVZ

Presumed *direction of neuron migration* from neuroepithelial sources.

Presumed *direction of axon growth* in brain fiber tracts.

Telencephalic choroid plexus

TELENCEPHALIC SUPERVENTRICLE (FUTURE LATERAL VENTRICLE)

Cortical (hippocampal) NEP and SVZ

HIPPO-CAMPUS

Future Ammon's horn

Diencephalic choroid plexus

Ammonic migration

Dentate migration

Corticoganglionic NEP and SVZ

Pioneer corticofugal fibers enter internal capsule in STF4

Fornical glioepithelium

Choroid plexus stem cells

Fornix

Stria medullaris

Dorsolateral nucleus

Posterior ganglionic NEP and SVZ

Lateral migratory stream percolates through STF4

Periventricular complex

Dorsomedial nucleus

Ventral anterior nucleus

Caudate

THALAMUS

Stria terminalis

THIRD VENTRICLE

Thalamic glioepithelium/ ependyma (G/EP)

Ventral complex

Ventral complex

Reticular nucleus

Thalamocortical fibers funnel into cortex via the internal capsule

Putamen

Future insular cortex

Pioneer corticofugal fibers

Subthalamic G/EP

BASAL GANGLIA

Internal capsule

Globus pallidus

External segment

Lateral migratory stream infiltrates external capsule?

SUBTHALAMUS
Forel's fields

Internal segment

Putamen

FUTURE TEMPORAL LOBE

Hypothalamic G/EP

Dorsomedial nucleus

Ventral striatum

External capsule

HYPOTHALAMUS

Strionuclear glioepithelium

PLATE 46A
CR 60 mm, GW 12.5, Y1-59
Frontal
Section 399

ENTRY/EXIT ZONE IN
PARACENTRAL CORTEX

See nearby sections in
Plates 29 to 30A and B.

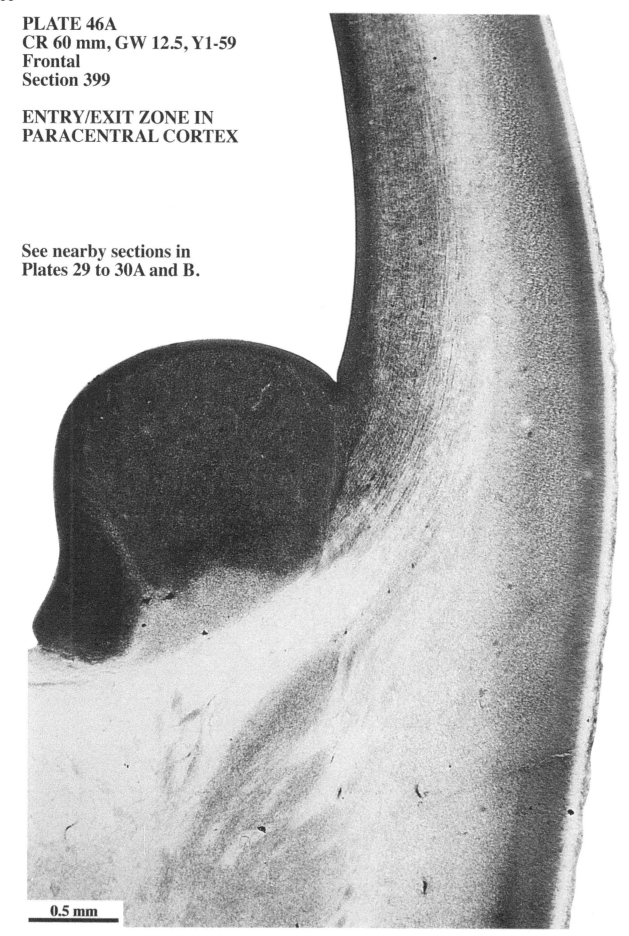

0.5 mm

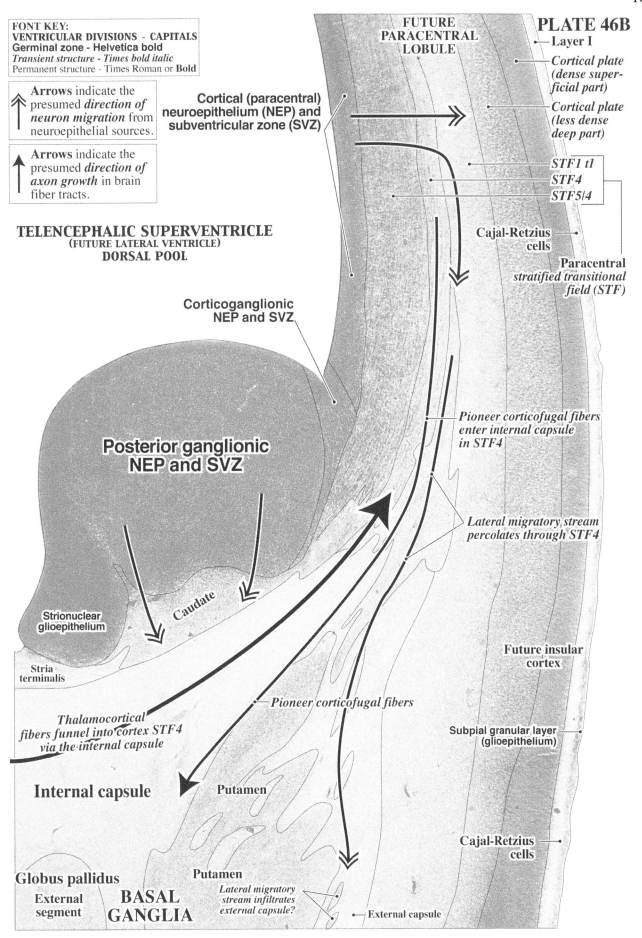

FONT KEY:
VENTRICULAR DIVISIONS - CAPITALS
Germinal zone - Helvetica bold
Transient structure - Times bold italic
Permanent structure - Times Roman or **Bold**

Arrows indicate the presumed *direction of neuron migration* from neuroepithelial sources.

Arrows indicate the presumed *direction of axon growth* in brain fiber tracts.

TELENCEPHALIC SUPERVENTRICLE
(FUTURE LATERAL VENTRICLE)
DORSAL POOL

Cortical (paracentral) neuroepithelium (NEP) and subventricular zone (SVZ)

Corticoganglionic NEP and SVZ

Posterior ganglionic NEP and SVZ

Strionuclear glioepithelium

Stria terminalis

Caudate

Thalamocortical fibers funnel into cortex STF4 via the internal capsule

Internal capsule

Putamen

Globus pallidus

External segment

Putamen

BASAL GANGLIA

Lateral migratory stream infiltrates external capsule?

External capsule

FUTURE PARACENTRAL LOBULE

PLATE 46B
Layer I
Cortical plate (dense superficial part)
Cortical plate (less dense deep part)

STF1 t1
STF4
STF5/4

Cajal-Retzius cells

Paracentral *stratified transitional field (STF)*

Pioneer corticofugal fibers enter internal capsule in STF4

Lateral migratory stream percolates through STF4

Future insular cortex

Pioneer corticofugal fibers

Subpial granular layer (glioepithelium)

Cajal-Retzius cells

DIENCEPHALON, BASAL GANGLIA,
and BASAL TELENCEPHALON

1 mm

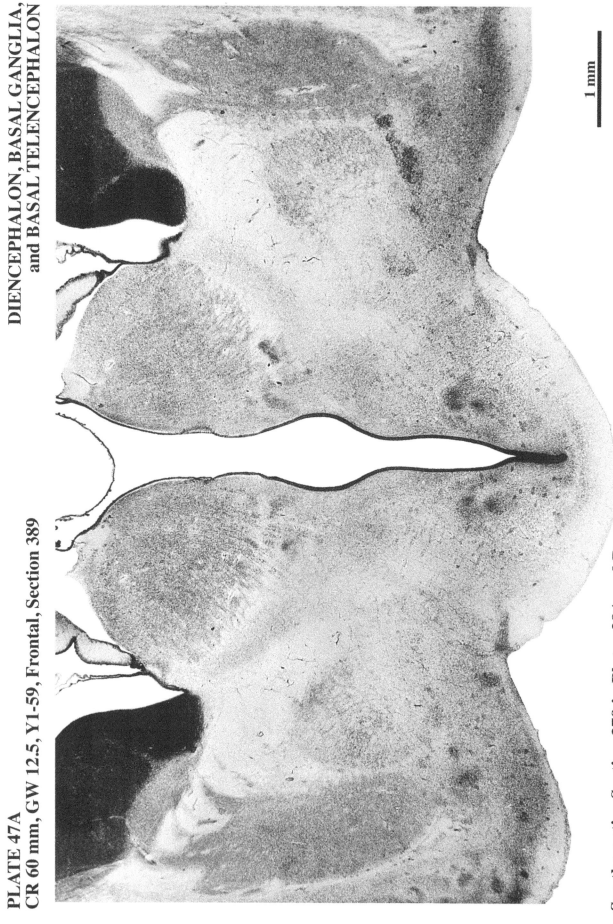

PLATE 47A
CR 60 mm, GW 12.5, Y1-59, Frontal, Section 389

See the entire Section 379 in Plates 29A and B.

109

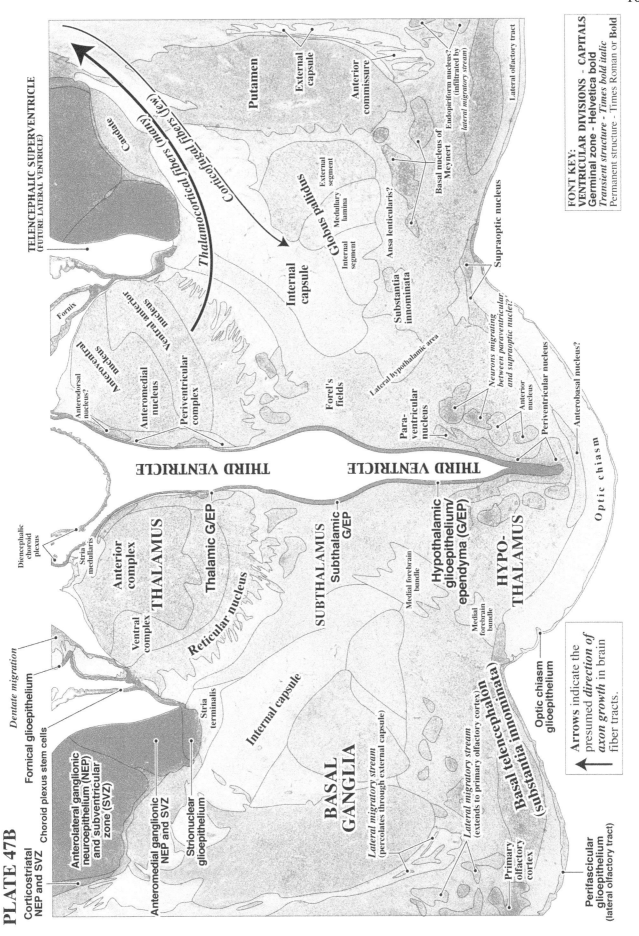

PLATE 47B

110

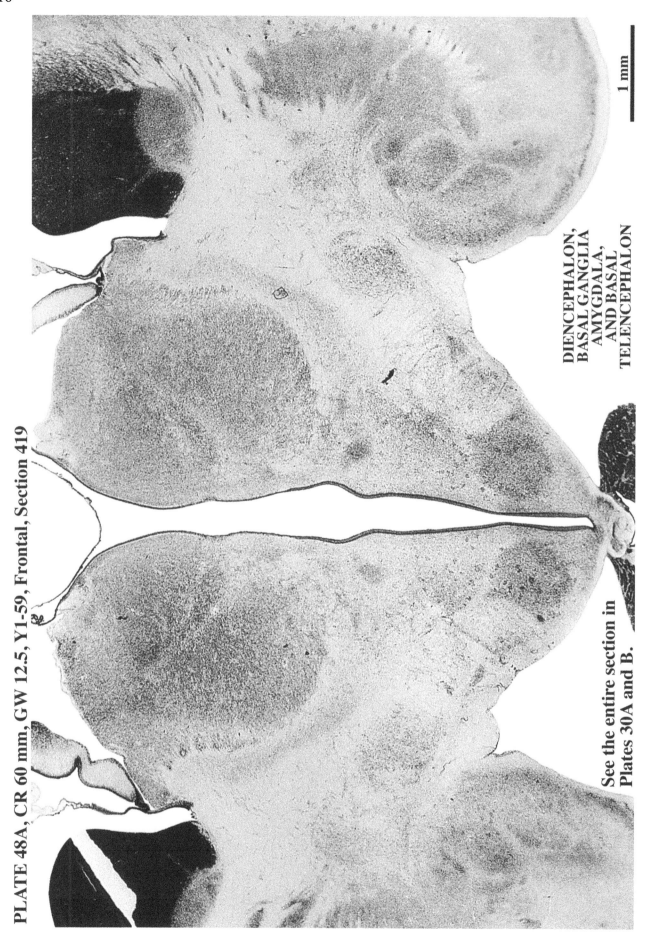

PLATE 48A, CR 60 mm, GW 12.5, Y1-59, Frontal, Section 419

DIENCEPHALON,
BASAL GANGLIA,
AMYGDALA,
AND BASAL
TELENCEPHALON

See the entire section in
Plates 30A and B.

1 mm

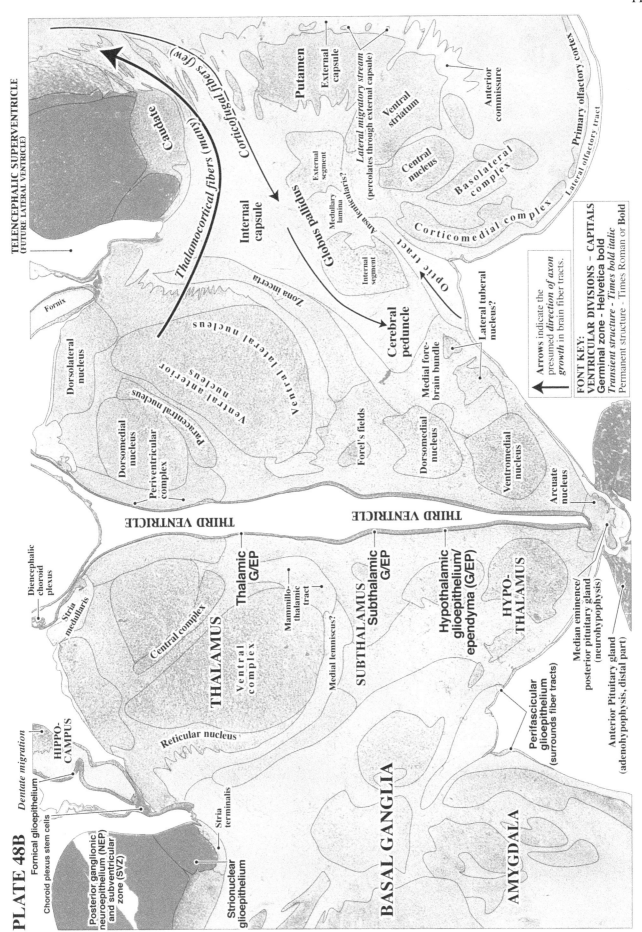

PLATE 48B

Dentate migration

Fornical glioepithelium

Choroid plexus stem cells

Posterior ganglionic neuroepithelium (NEP) and subventricular zone (SVZ)

Strionuclear glioepithelium

HIPPO-CAMPUS

Diencephalic choroid plexus

Stria medullaris

Fornix

Central complex

THALAMUS

Ventral complex

Reticular nucleus

Thalamic G/EP

Mammillo-thalamic tract

Medial lemniscus?

SUBTHALAMUS
Subthalamic G/EP

Stria terminalis

Perifascicular glioepithelium (surrounds fiber tracts)

BASAL GANGLIA

Hypothalamic glioepithelium/ependyma (G/EP)

HYPO-THALAMUS

Median eminence/posterior pituitary gland (neurohypophysis)

Anterior Pituitary gland (adenohypophysis, distal part)

AMYGDALA

THIRD VENTRICLE

THIRD VENTRICLE

TELENCEPHALIC SUPERVENTRICLE
(FUTURE LATERAL VENTRICLE)

Caudate

Thalamocortical fibers (many)

Corticofugal fibers (few)

Internal capsule

Globus pallidus

Putamen

External capsule

External segment

Medullary lamina

Ansa lenticularis?

Internal segment

Lateral migratory stream (percolates through external capsule)

Ventral striatum

Central nucleus

Anterior commissure

Basolateral complex

Corticomedial complex

Primary olfactory cortex

Lateral olfactory tract

Dorsolateral nucleus

Dorsomedial nucleus

Paracentral nucleus

Periventricular complex

Ventral anterior nucleus

Ventral lateral nucleus

Zona incerta

Cerebral peduncle

Optic tract

Forel's fields

Dorsomedial nucleus

Ventromedial nucleus

Arcuate nucleus

Medial fore-brain bundle

Lateral tuberal nucleus?

FONT KEY:
VENTRICULAR DIVISIONS – CAPITALS
Germinal zone - **Helvetica bold**
Transient structure - Times bold italic
Permanent structure - Times Roman or **Bold**

Arrows indicate the presumed *direction of axon growth* in brain fiber tracts.

PLATE 49A
CR 60 mm, GW 12.5, Y1-59, Frontal, Section 449

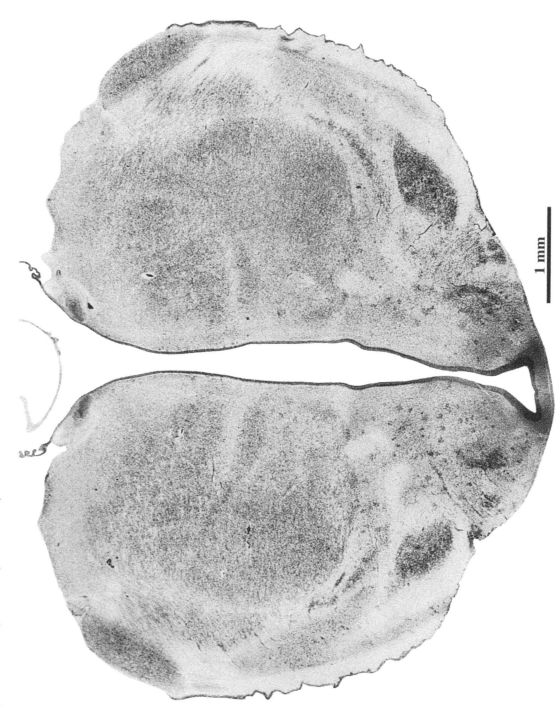

1 mm

See nearby complete sections Plates 31 to 32A and B.

PLATE 49B

FONT KEY:
VENTRICULAR DIVISIONS - CAPITALS
Germinal zone - **Helvetica bold**
Transient structure - Times bold italic
Permanent structure - Times Roman or **Bold**

ABBREVIATIONS:
GEP - Glioepithelium
G/EP - Glioepithelium/ependyma
NEP - Neuroepithelium

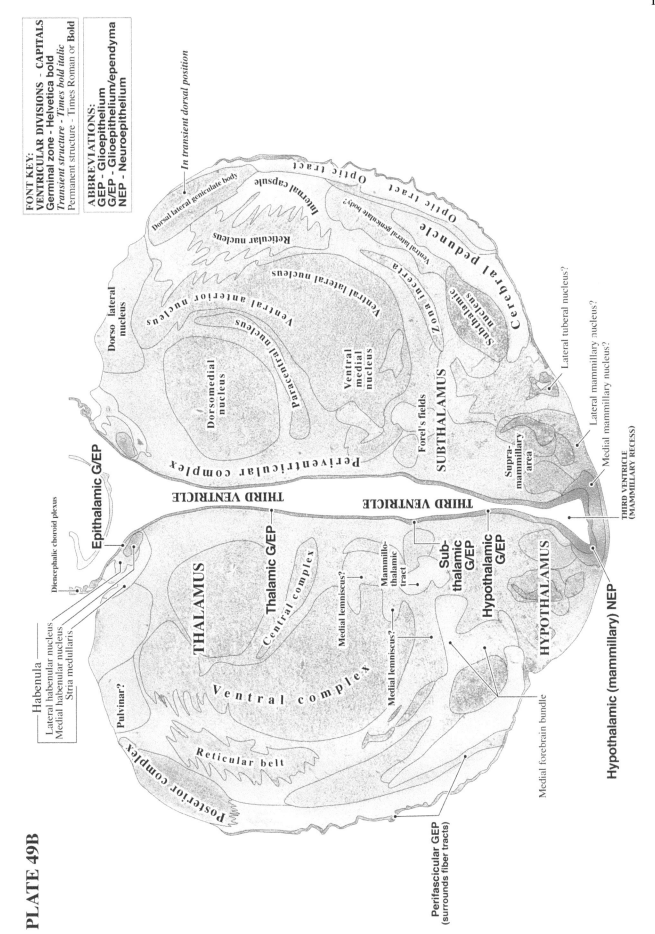

In transient dorsal position

Dorsal lateral geniculate body

Optic tract

Internal capsule

Reticular nucleus

Dorso lateral nucleus

Ventral anterior nucleus

Ventral lateral nucleus

Ventral lateral geniculate body?

Optic tract

Zona incerta

Dorsomedial nucleus

Paracentral nucleus

Ventral medial nucleus

Subthalamic nucleus

Cerebral peduncle

Forel's fields

SUBTHALAMUS

Lateral tuberal nucleus?

Supra-mammillary area

Lateral mammillary nucleus?

Medial mammillary nucleus?

THIRD VENTRICLE (MAMMILLARY RECESS)

Epithalamic G/EP

Periventricular complex

THIRD VENTRICLE

THIRD VENTRICLE

Diencephalic choroid plexus

Habenula
Lateral habenular nucleus
Medial habenular nucleus
Stria medullaris

THALAMUS

Thalamic G/EP

Central complex

Mammillo-thalamic tract

Sub-thalamic G/EP

Hypothalamic G/EP

HYPOTHALAMUS

Pulvinar?

Ventral complex

Medial lemniscus?

Medial lemniscus?

Medial lemniscus?

Reticular belt

Posterior complex

Medial forebrain bundle

Hypothalamic (mammillary) NEP

Perifascicular GEP (surrounds fiber tracts)

DIENCEPHALON

PLATE 50A
CR 60 mm, GW 12.5, Y1-59, Frontal, Section 469

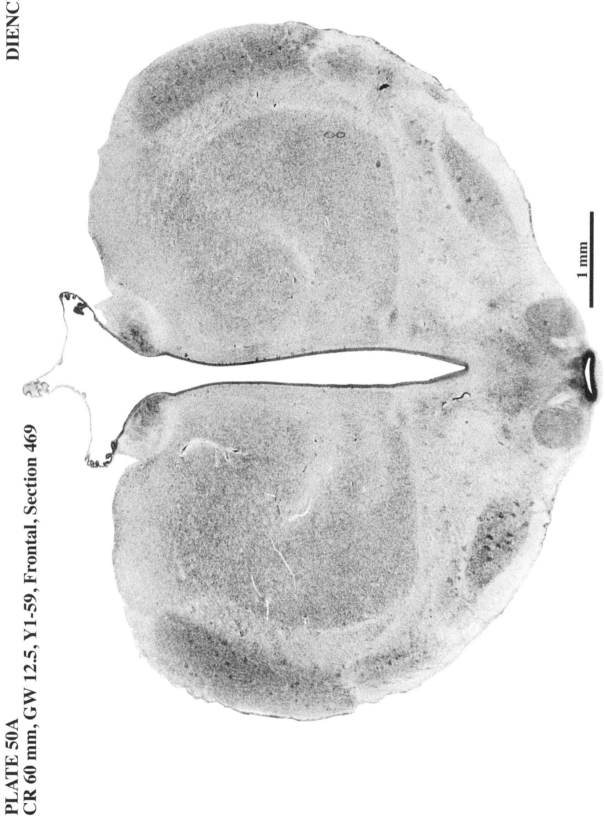

1 mm

See nearby complete sections in Plates 32 to 33A and B.

PLATE 50B

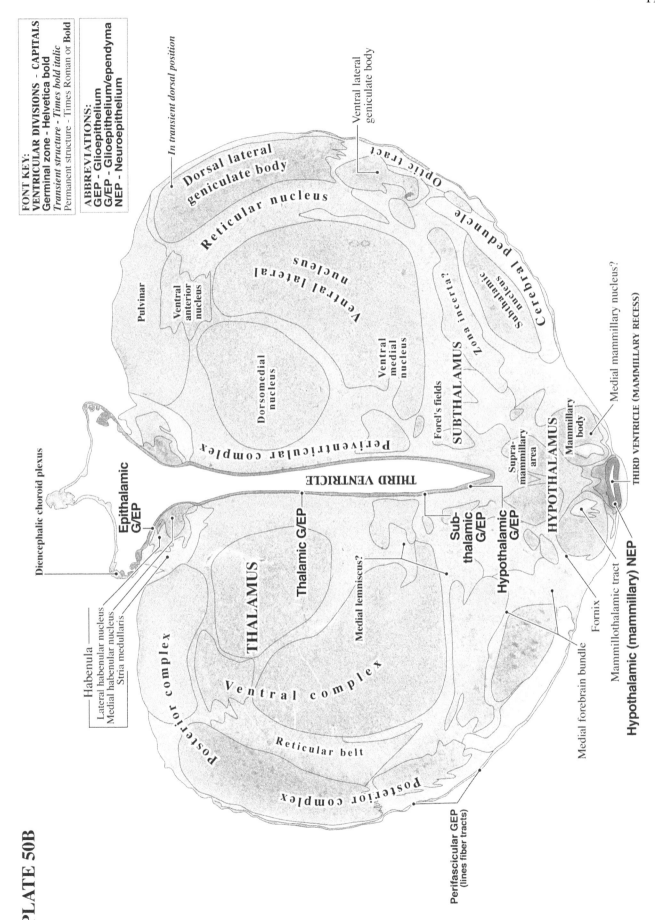

In transient dorsal position

Ventral lateral
geniculate body

**Dorsal lateral
geniculate body**

Optic tract

Reticular nucleus

Cerebral peduncle

Pulvinar

Ventral
anterior
nucleus

*Ventral lateral
nucleus*

*Subthalamic
nucleus*

Dorsomedial
nucleus

Ventral
medial
nucleus

Zona incerta?

SUBTHALAMUS

Forel's fields

Medial mammillary nucleus?

THIRD VENTRICLE (MAMMILLARY RECESS)

Periventricular complex

Supra-
mammillary
area

Mammillary
body

THIRD VENTRICLE

HYPOTHALAMUS

Diencephalic choroid plexus

**Epithalamic
G/EP**

Thalamic G/EP

**Sub-
thalamic
G/EP**

**Hypothalamic
G/EP**

Habenula
Lateral habenular nucleus
Medial habenular nucleus
Stria medullaris

THALAMUS

Medial lemniscus?

Hypothalamic (mammillary) NEP

Fornix

Mammillothalamic tract

Posterior complex

Ventral complex

Medial forebrain bundle

Reticular belt

Posterior complex

Perifascicular GEP
(lines fiber tracts)

DIENCEPHALON

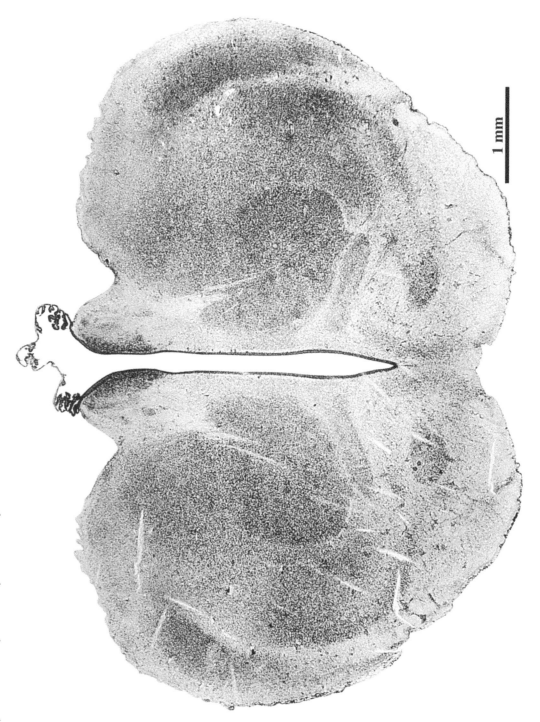

PLATE 51A
CR 60 mm, GW 12.5, Y1-59, Frontal, Section 499

1 mm

See the entire section 500 in Plates 34A and B.

PLATE 51B

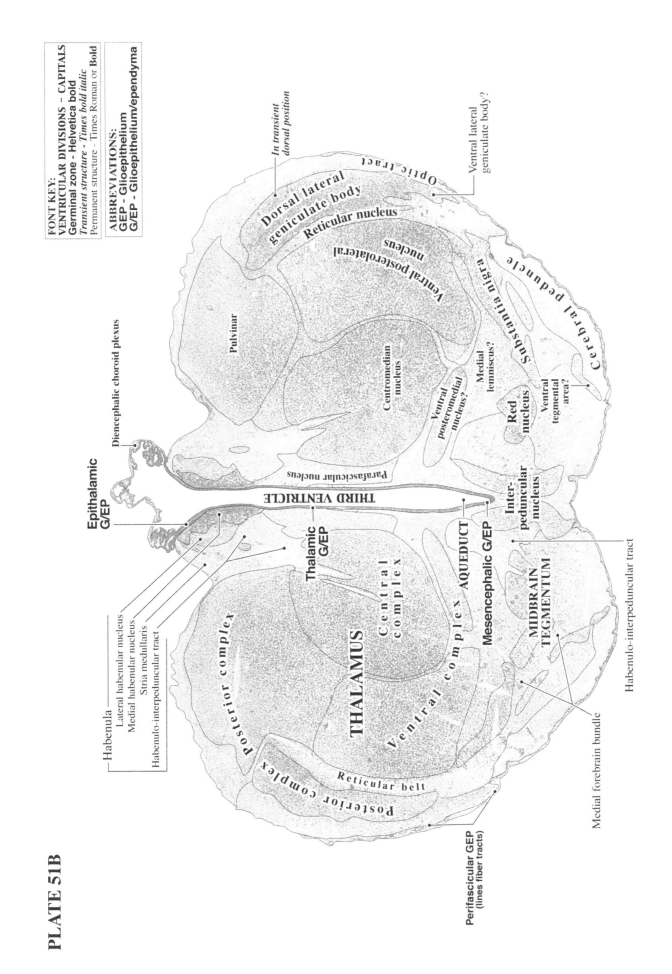

FONT KEY:
VENTRICULAR DIVISIONS – **CAPITALS**
Germinal zone - Helvetica bold
Transient structure - Times bold italic
Permanent structure - Times Roman or **Bold**

ABBREVIATIONS:
GEP - Glioepithelium
G/EP - Glioepithelium/ependyma

In transient dorsal position

Dorsal lateral geniculate body

Optic tract

Reticular nucleus

Ventral posterolateral nucleus

Ventral lateral geniculate body?

Pulvinar

Cerebral peduncle

Substantia nigra

Diencephalic choroid plexus

Centromedian nucleus

Medial lemniscus?

Ventral posteromedial nucleus?

Red nucleus

Ventral tegmental area?

Epithalamic G/EP

Parafascicular nucleus

THIRD VENTRICLE

Interpeduncular nucleus

Habenula
Lateral habenular nucleus
Medial habenular nucleus
Stria medullaris
Habenulo-interpeduncular tract

Thalamic G/EP

Central complex

AQUEDUCT

Mesencephalic G/EP

Habenulo-interpeduncular tract

Posterior complex

THALAMUS

Ventral complex

MIDBRAIN TEGMENTUM

Posterior complex

Reticular belt

Medial forebrain bundle

Perifascicular GEP (lines fiber tracts)

118

DIENCEPHALON AND MIDBRAIN TEGMENTUM

PLATE 52A
CR 60 mm, GW 12.5, Y1-59, Frontal, Section 529

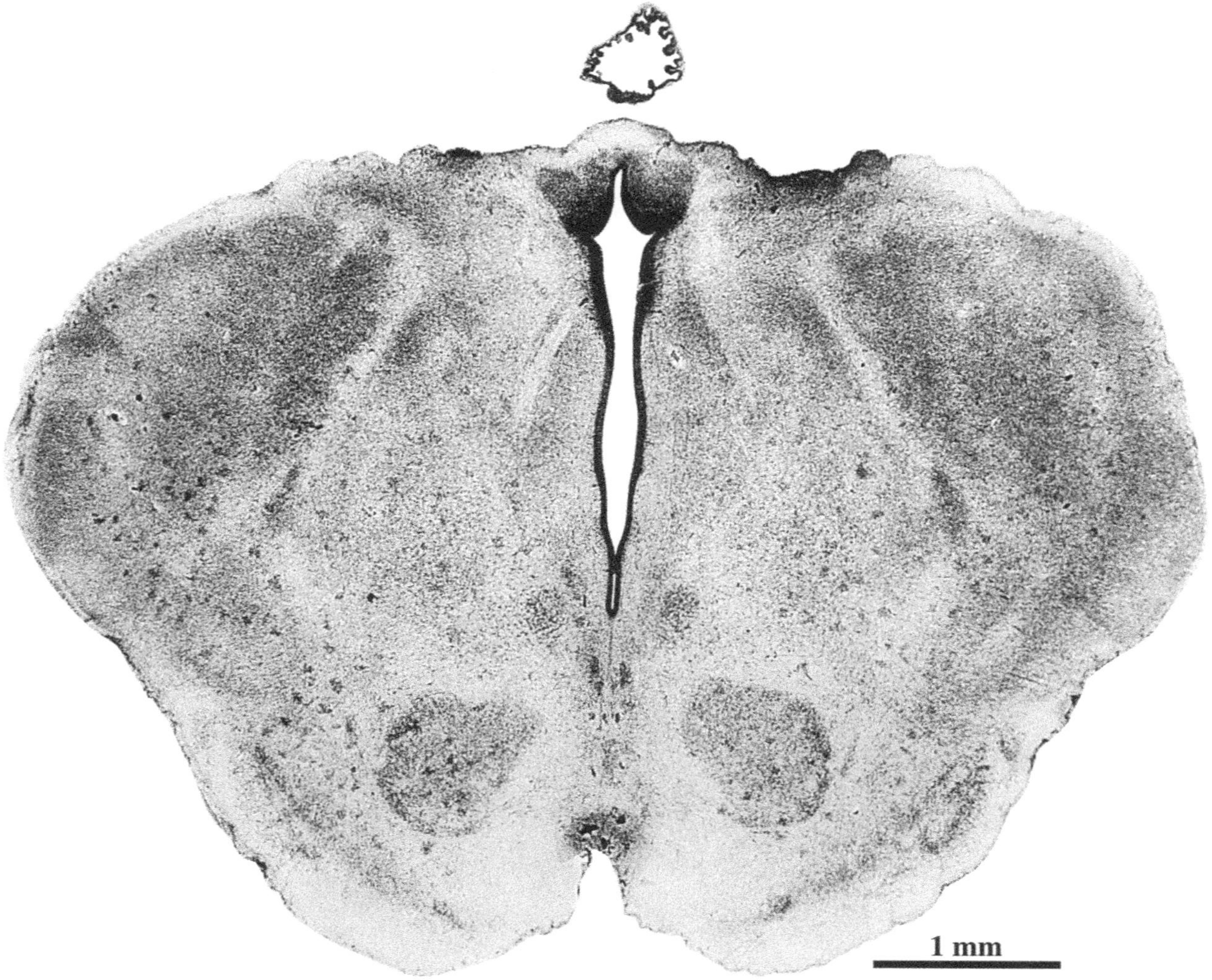

1 mm

See the entire section in Plates 35A and B.

PLATE 52B

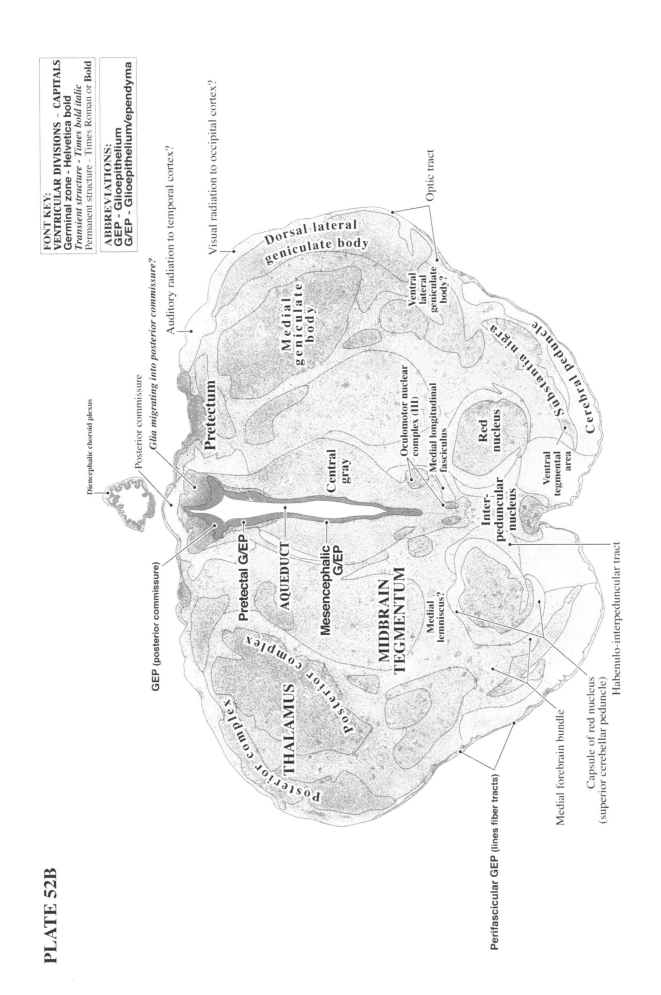

Diencephalic choroid plexus

Posterior commissure

GEP (posterior commissure)

Glia migrating into posterior commissure?

Auditory radiation to temporal cortex?

Visual radiation to occipital cortex?

Optic tract

Dorsal-lateral geniculate body

Medial geniculate body

Ventral lateral geniculate body?

Pretectum

Substantia nigra

Cerebral peduncle

Central gray

Oculomotor nuclear complex (III)

Medial longitudinal fasciculus

Red nucleus

Ventral tegmental area

Pretectal G/EP

AQUEDUCT

Mesencephalic G/EP

Inter- peduncular nucleus

MIDBRAIN TEGMENTUM

Medial lemniscus?

Posterior complex

THALAMUS

Perifascicular GEP (lines fiber tracts)

Medial forebrain bundle

Capsule of red nucleus (superior cerebellar peduncle)

Habenulo-interpeduncular tract

MIDBRAIN AND PONS

PLATE 53A
CR 60 mm, GW 12.5, Y1-59,
Frontal
Section 549

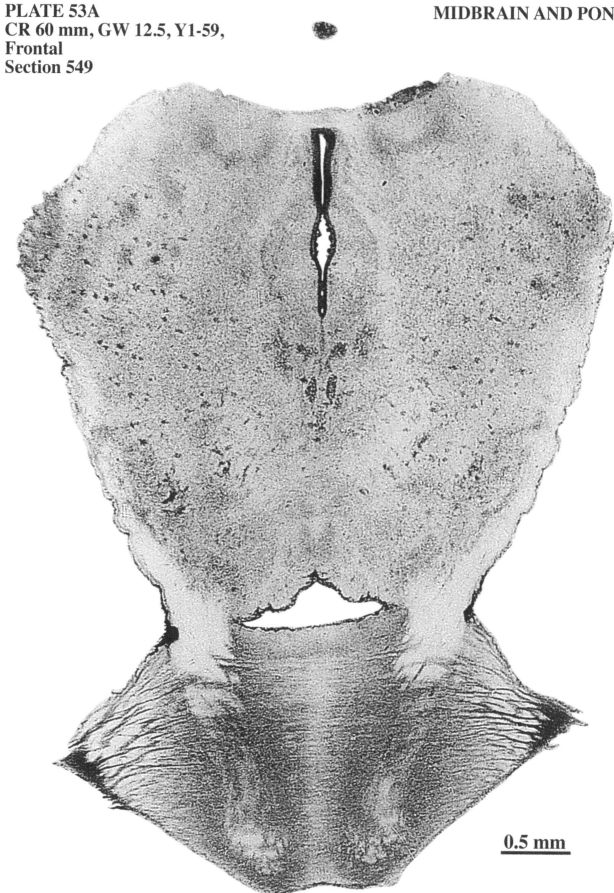

0.5 mm

See the entire section in Plates 36A and B.

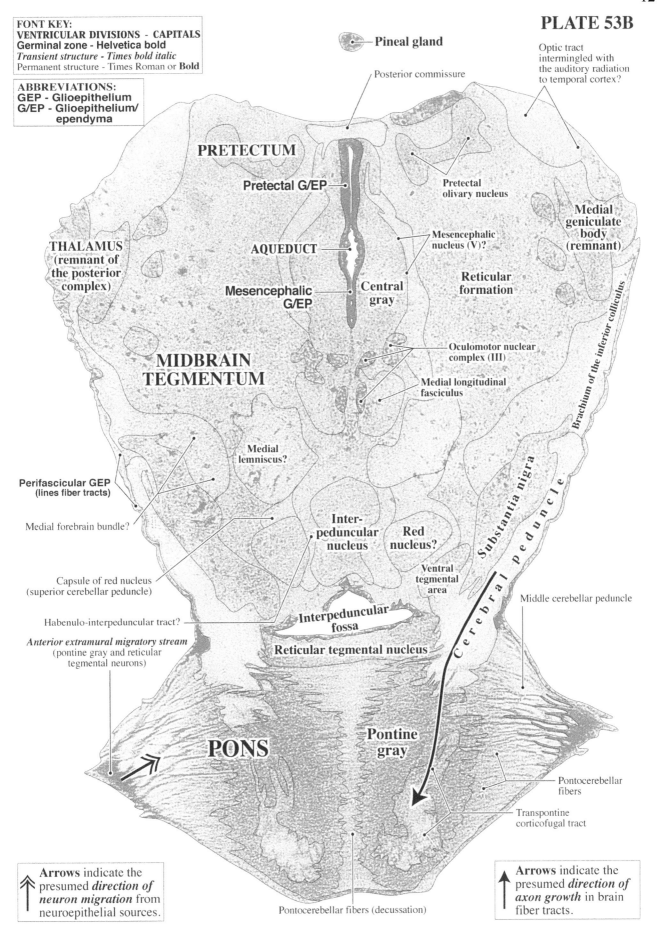

Pineal gland

Posterior commissure

Optic tract
intermingled with
the auditory radiation
to temporal cortex?

PRETECTUM

Pretectal G/EP

Pretectal
olivary nucleus

**Medial
geniculate
body
(remnant)**

Mesencephalic
nucleus (V)?

**THALAMUS
(remnant of
the posterior
complex)**

AQUEDUCT

Mesencephalic
G/EP

Central
gray

**Reticular
formation**

Oculomotor nuclear
complex (III)

**MIDBRAIN
TEGMENTUM**

Medial longitudinal
fasciculus

Brachium of the inferior colliculus

Medial
lemniscus?

Perifascicular GEP
(lines fiber tracts)

Medial forebrain bundle?

Inter-
peduncular
nucleus

Red
nucleus?

Substantia nigra

Capsule of red nucleus
(superior cerebellar peduncle)

Ventral
tegmental
area

Cerebral peduncle

Middle cerebellar peduncle

Habenulo-interpeduncular tract?

Interpeduncular
fossa

Anterior extramural migratory stream
(pontine gray and reticular
tegmental neurons)

Reticular tegmental nucleus

PONS

**Pontine
gray**

Pontocerebellar
fibers

Transpontine
corticofugal tract

Pontocerebellar fibers (decussation)

Arrows indicate the
presumed *direction of
neuron migration* from
neuroepithelial sources.

Arrows indicate the
presumed *direction of
axon growth* in brain
fiber tracts.

PLATE 54A
CR 60 mm, GW 12.5, Y1-59
Frontal
Section 569

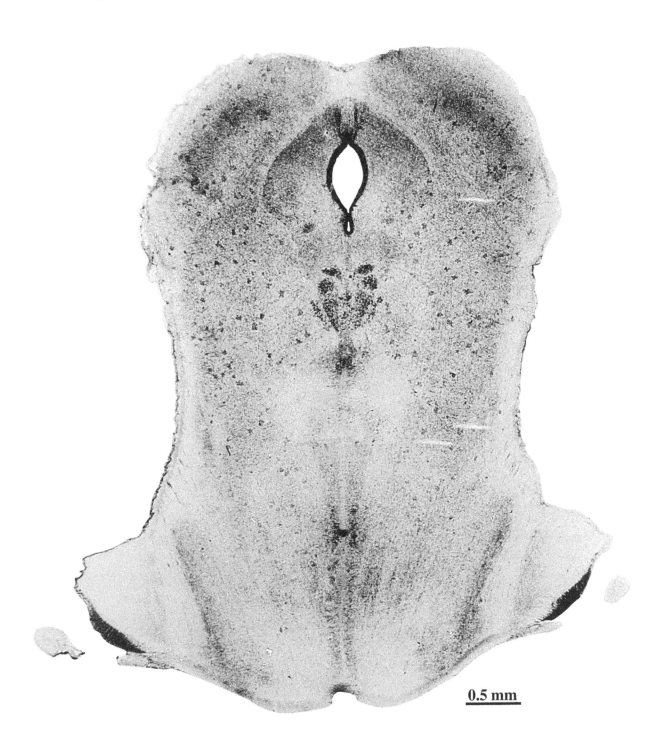

0.5 mm

See the entire section Plates 36A and B.

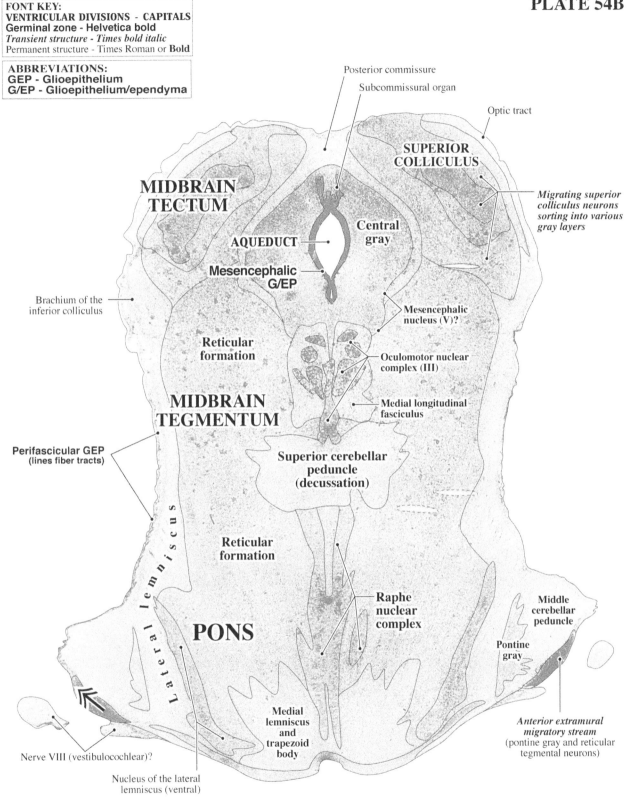

Posterior commissure

Subcommissural organ

Optic tract

SUPERIOR COLLICULUS

MIDBRAIN TECTUM

Migrating superior colliculus neurons sorting into various gray layers

Central gray

AQUEDUCT

Mesencephalic G/EP

Brachium of the inferior colliculus

Mesencephalic nucleus (V)?

Reticular formation

Oculomotor nuclear complex (III)

MIDBRAIN TEGMENTUM

Medial longitudinal fasciculus

Perifascicular GEP (lines fiber tracts)

Superior cerebellar peduncle (decussation)

L a t e r a l   l e m n i s c u s

Reticular formation

PONS

Raphe nuclear complex

Middle cerebellar peduncle

Pontine gray

Nerve VIII (vestibulocochlear)?

Medial lemniscus and trapezoid body

Anterior extramural migratory stream (pontine gray and reticular tegmental neurons)

Nucleus of the lateral lemniscus (ventral)

Dashed lines indicate staining and/or sectioning artifacts.

Arrows indicate the presumed *direction of neuron migration* from neuroepithelial sources.

124

PLATE 55A
CR 60 mm, GW 12.5, Y1-59
Frontal
Section 589

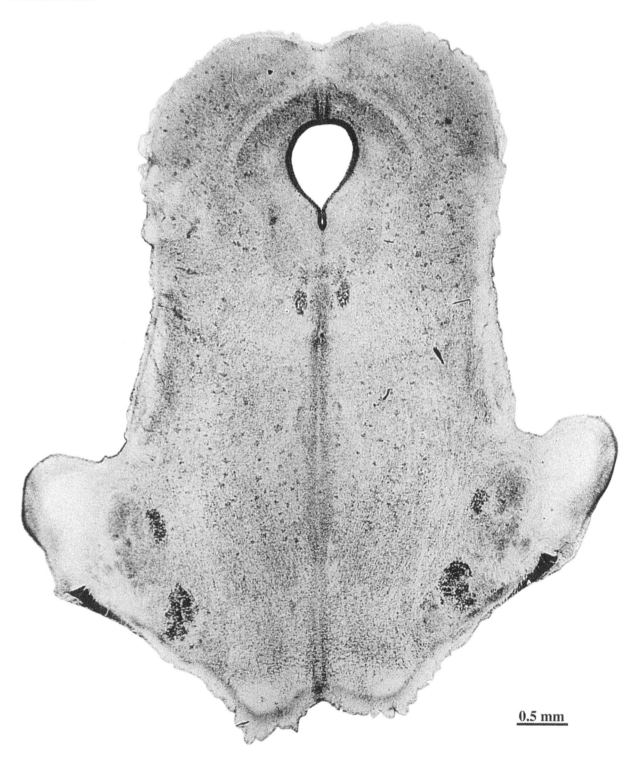

0.5 mm

See a nearby complete section in Plates 37A and B.

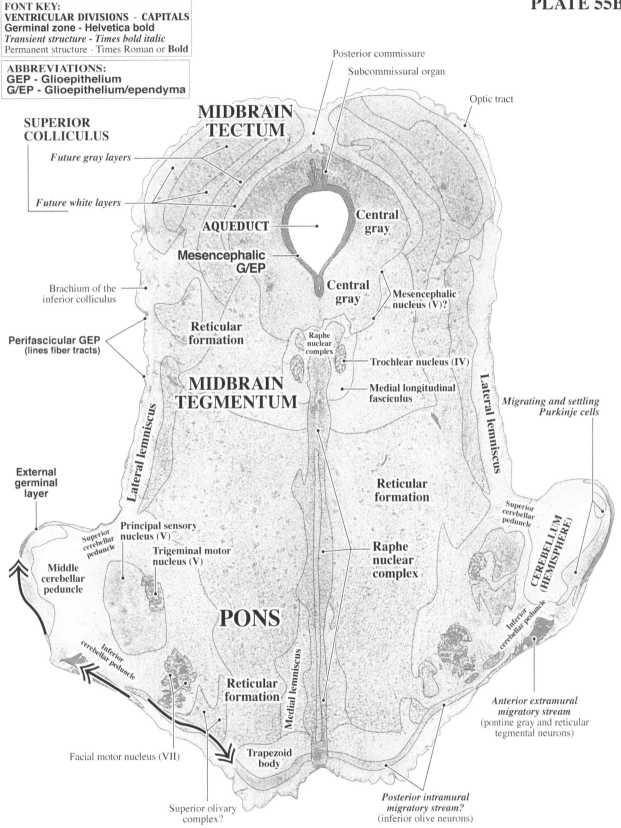

Posterior commissure

Subcommissural organ

Optic tract

MIDBRAIN TECTUM

SUPERIOR COLLICULUS

Future gray layers

Future white layers

Central gray

AQUEDUCT

Mesencephalic G/EP

Central gray

Mesencephalic nucleus (V)?

Brachium of the inferior colliculus

Reticular formation

Raphe nuclear complex

Trochlear nucleus (IV)

Perifascicular GEP (lines fiber tracts)

MIDBRAIN TEGMENTUM

Medial longitudinal fasciculus

Lateral lemniscus

Migrating and settling Purkinje cells

Lateral lemniscus

Reticular formation

Superior cerebellar peduncle

External germinal layer

Superior cerebellar peduncle

Principal sensory nucleus (V)

Raphe nuclear complex

CEREBELLUM (HEMISPHERE)

Middle cerebellar peduncle

Trigeminal motor nucleus (V)

Inferior cerebellar peduncle

Inferior cerebellar peduncle

PONS

Medial lemniscus

Reticular formation

Anterior extramural migratory stream (pontine gray and reticular tegmental neurons)

Facial motor nucleus (VII)

Trapezoid body

Superior olivary complex?

Posterior intramural migratory stream? (inferior olive neurons)

Arrows indicate the presumed *direction of neuron migration* from neuroepithelial sources.

PLATE 56A
CR 60 mm, GW 12.5, Y1-59
Frontal
Section 599

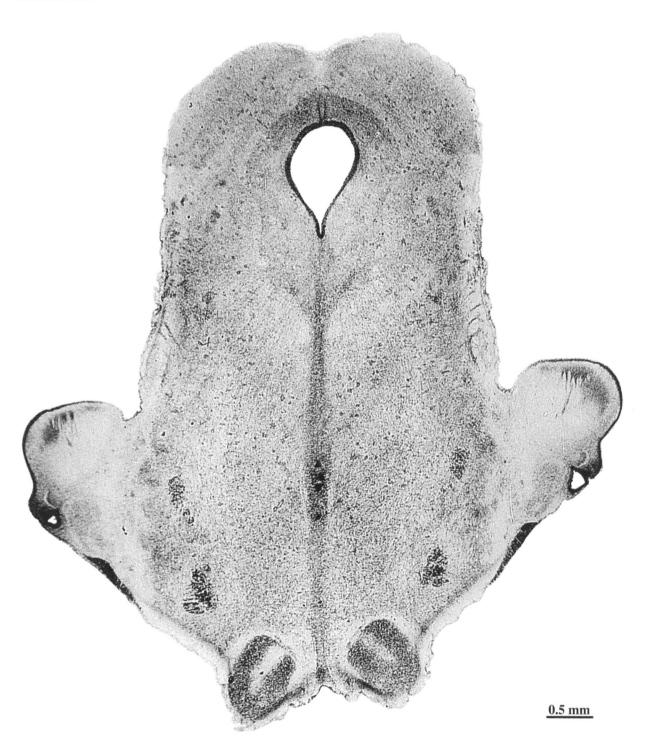

0.5 mm

See the entire section in Plates 37A and B.

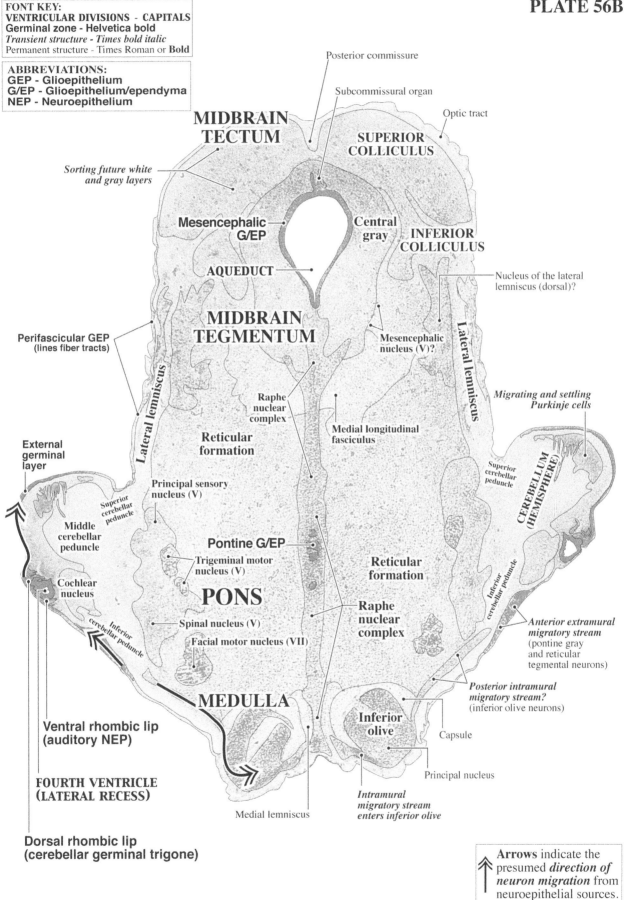

Posterior commissure

Subcommissural organ

Optic tract

MIDBRAIN TECTUM

SUPERIOR COLLICULUS

Sorting future white and gray layers

Mesencephalic G/EP

Central gray

INFERIOR COLLICULUS

AQUEDUCT

Nucleus of the lateral lemniscus (dorsal)?

MIDBRAIN TEGMENTUM

Mesencephalic nucleus (V)?

Lateral lemniscus

Perifascicular GEP (lines fiber tracts)

Raphe nuclear complex

Medial longitudinal fasciculus

Reticular formation

Migrating and settling Purkinje cells

Lateral lemniscus

Superior cerebellar peduncle

CEREBELLUM (HEMISPHERE)

External germinal layer

Principal sensory nucleus (V)

Superior cerebellar peduncle

Middle cerebellar peduncle

Pontine G/EP

Reticular formation

Inferior cerebellar peduncle

Cochlear nucleus

Trigeminal motor nucleus (V)

Inferior cerebellar peduncle

PONS

Spinal nucleus (V)

Raphe nuclear complex

Anterior extramural migratory stream (pontine gray and reticular tegmental neurons)

Facial motor nucleus (VII)

Posterior intramural migratory stream? (inferior olive neurons)

MEDULLA

Inferior olive

Capsule

Ventral rhombic lip (auditory NEP)

Principal nucleus

FOURTH VENTRICLE (LATERAL RECESS)

Intramural migratory stream enters inferior olive

Medial lemniscus

Dorsal rhombic lip (cerebellar germinal trigone)

Arrows indicate the presumed *direction of neuron migration* from neuroepithelial sources.

9 781032 185668